Internet links

There are lots of useful websites where you can find out more about biology. We have created links to some of the best sites on the Usborne Quicklinks Website. To visit the sites, go to **www.usborne-quicklinks.com** and type the keywords "biology dictionary". Here are some of the things you can do on the Internet:

• Try online puzzles, games, quizzes and experiments
• Email a life sciences question to a biologist
• Dissect a virtual frog online

Internet safety

The websites recommended in Usborne Quicklinks are regularly reviewed. However, the content of a website may change at any time and Usborne Publishing is not responsible for the content of websites other than its own. We recommend that children are supervised while on the Internet.

The Usborne
Illustrated
Dictionary
of
Biology

Corinne Stockley
Revision editor: Kirsteen Rogers
Designers: Karen Tomlins and Verinder Bhachu
Illustrators: Kuo Kang Chen and Guy Smith

Scientific advisors:
Dr. Margaret Rostron and Dr. John Rostron

ABOUT BIOLOGY

Biology is the study of living things. It examines the structures and internal systems of different organisms and how these operate to sustain individual life, as well as looking at the complex web of relationships between organisms which ensure new life is created and maintained. In this book, biology is divided into six colour-coded sections. The areas covered by these sections are explained below.

Ecology and living things

Looks at the complex relationships between all living things, and their basic cellular structure.

Zoology (humans)

Covers all the major terms of human biology. In many cases, these also apply to the vertebrates in general (see page 113).

Botany

Covers the plant kingdom. Introduces the different types of plant, their main characteristics, internal structures and systems.

Reproduction and genetics

Examines the different types of reproduction and introduces the branch of biology known as genetics.

Zoology (animals)

Examines the component parts, systems and behaviour typical of the major animal groups other than humans.

General biology information

Covers subjects which relate to all living things. Includes tables of general information and classification charts.

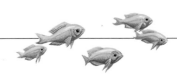

CONTENTS

LIVING THINGS AND THEIR ENVIRONMENT

The world can be divided into a number of different regions, each with its own characteristic plants and animals. All the plants and animals have become adapted to their own surroundings, or **environment** (see **adaptive radiation**, page 9), and their lives are linked in a complex web of interdependence. The environment is influenced by many different factors, e.g. temperature, water and light (**climatic factors**), the physical and chemical properties of the soil (**edaphic factors**), and the activities of living things (**biotic factors**). The study of the relationships between plants, animals and the environment is called **ecology**.

This tree frog's toes have adapted to help it cling to bark.

Biosphere

The layer of the Earth (including the oceans and the atmosphere) which is inhabited by living things. The biosphere's boundaries are the upper atmosphere (above) and the first layers of uninhabited rock (below).

Biomes

The main ecological regions into which the land surface can be divided. Each has its own characteristic seasons, day length, rainfall pattern and maximum and minimum temperatures. The major biomes (see map, above right) are **tundra**, **coniferous forest**, **deciduous forest**, **tropical forest**, **temperate grassland**, **savanna** (tropical grassland) and **desert**. Most are named after the dominant vegetation, since this determines all other living things found there. Each biome is a giant **habitat** (**macrohabitat**). Human activity, e.g. deforestation in tropical forests, has begun to have harmful effects on the habitats which exist within many biomes.

Deforestation is a threat to huge areas of tropical rainforest, and the plants and animals that live there.

Map showing main world biomes

Key to biomes on map above

Tundra
Very cold and windy. Commonest plants: **lichens*** and small shrubs. Animals include musk ox.

Coniferous forest
Low temperatures all year. Dominant plants: **conifers***, e.g. spruce. Commonest large animals: deer.

Deciduous forest
Summers warm, winters cold. Dominant plants: **deciduous*** trees, e.g. beech. Many animals, e.g. foxes.

Tropical forest
High temperatures all year, heavy rainfall. Great variety of plants and animals, e.g. exotic birds.

Desert
High temperatures (cold at night), very low rainfall. Typical plants: cacti. Animals include jerboas, scorpions.

Temperate grassland
Open grassy plains. Hot summers and cold winters. Main plants: grasses. Animals include prairie dogs.

Savanna
Main plants: grasses, but rainfall enough for trees. Typical animals: giraffe.

Scrubland (**maquis**)

Mountains

Ice

* **Conifers**, 112; **Deciduous**, 8; **Lichens**, 114 (**Mutualists**).

Habitat

The natural home of a group of living things or a single living thing. Small habitats can be found within large habitats, e.g. a waterhole in the **savanna biome**. Very small specialized habitats are called **microhabitats**, e.g. a rotting acacia tree.

Community

The group of plants and animals found in one **habitat**. They all interact with each other and their environment.

Waterhole and acacia tree habitats in savanna

Ecosystem

The **community** of plants and animals in a given **habitat**, together with the non-living parts of the environment (e.g. air or water). An ecosystem is a self-contained unit, i.e. the plants and animals interact to produce all the material they need (see also pages 6-7).

Ecosystem includes environment, e.g. air and water.

The **community** includes antelopes and ostriches.

Ecological succession

A process which occurs whenever a new area of land is colonized, e.g. a forest floor after a fire, a farm field which is left uncultivated or a demolition site which remains unused. Over the years, different types of plants (and the animals which go with them) will succeed each other, until a **climax community** is arrived at. This is a very stable **community**, one which will survive without change as long as the same conditions prevail (e.g. the climate).

Ecological succession in a disused field

Pioneer community (first **community**) of grasses, with insects, field mice, etc.

Successional community (intermediate **community**) of shrubs and bushes, with rabbits, thrushes, etc.

Climax community of **deciduous*** trees, e.g. oak and beech, with foxes, badgers, warblers, etc.

Ecological niche

The place held in an **ecosystem** by a plant or animal, e.g. what it eats and where it lives. **Gause's principle** states that no two species can occupy the same niche at the same time (if they tried, one species would die out or be driven away). For example, in the winter months, both the curlew and the ringed plover can be found living around the estuaries of Britain, and eating small creatures such as worms and snails. However, they actually occupy different niches. Curlews wade in the shallows, probing deep into the mud for food with their long beaks. Ringed plovers, by contrast, pick their food off the surface of the shore (their beaks are too short for probing). Hence both birds can survive in the same general area.

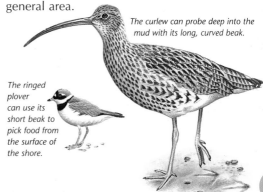

The curlew can probe deep into the mud with its long, curved beak.

The ringed plover can use its short beak to pick food from the surface of the shore.

* **Deciduous**, 8.

WITHIN AN ECOSYSTEM

An **ecosystem** consists of a group (**community***) of animals and plants which interact with each other and with their environment to produce a self-contained ecological unit.

Food web

The complex network of **food chains** in an ecosystem. A food chain is a linked series of living things, each of which is the food for the next in line. Plants make their food from non-living matter by **photosynthesis*** (they are **autotrophic**) and are always the first members of a chain. Animals cannot make their own food (they are **heterotrophic**) and so rely on the food-making activities of plants.

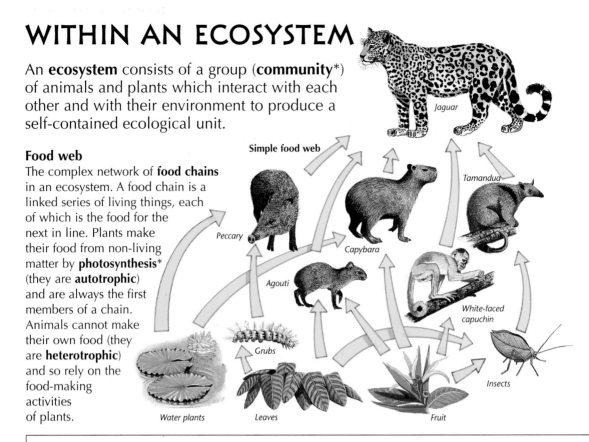

Simple food web

Jaguar

Peccary

Capybara

Tamandua

Agouti

White-faced capuchin

Grubs

Insects

Water plants

Leaves

Fruit

Carbon cycle

The constant circulation of the element carbon through living things and the atmosphere.

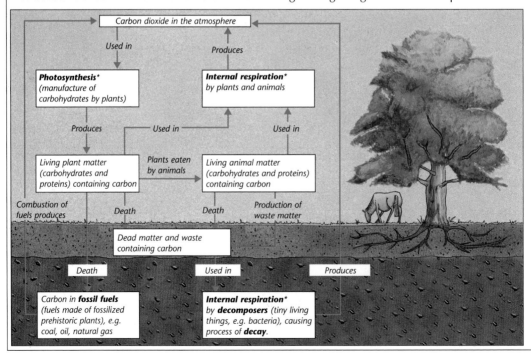

Carbon dioxide in the atmosphere

Used in

Produces

Photosynthesis* *(manufacture of carbohydrates by plants)*

Internal respiration* *by plants and animals*

Produces

Used in

Used in

Living plant matter *(carbohydrates and proteins) containing carbon*

Plants eaten by animals

Living animal matter *(carbohydrates and proteins) containing carbon*

Combustion of fuels produces

Death

Death

Production of waste matter

Dead matter and waste containing carbon

Death

Used in

Produces

Carbon in **fossil fuels** *(fuels made of fossilized prehistoric plants), e.g. coal, oil, natural gas*

Internal respiration* *by* **decomposers** *(tiny living things, e.g. bacteria), causing process of* **decay**.

* **Community**, 5; **Internal respiration**, 106; **Photosynthesis**, 26.

Trophic level or energy level

The level at which living things are positioned within a **food chain** (see **food web**). At each successive level, a great deal of the energy-giving food matter is lost. For example, a cow will break down well over half of the grass it eats (to provide energy). Hence only a small part of the original energy-giving material can be obtained from eating the cow (the part it used to build its own new tissue). This loss of energy means that the higher the trophic level, the fewer the number of animals, since they must eat progressively larger amounts of food to obtain enough energy. This principle is called the **pyramid of numbers**.

Pyramid of numbers

T4
T3
T2
T1

*Number of individuals at each **trophic level***

Pyramid of biomass

T4
T3
T2
T1

Total mass of individuals at each level (decrease is less extreme than left, since animals at higher levels tend to be larger).

Generalized food chain, showing trophic levels

Notes:

Producers – *green plants, which make their own food.* **Trophic level T1.**

1. **Omnivores**, *e.g. humans, eat plant and animal matter. They are thus placed on trophic level T2 at some times and on T3 (or T4) at others.*

Primary consumers or **first order consumers** – **herbivores** *(plant-eating animals), e.g. rabbits. Energy-giving material obtained directly from* **producers**. **Trophic level T2.**

2. *Many carnivores, e.g. foxes, will eat both herbivores and smaller carnivores. They are thus on trophic level T3 at some times and on T4 at others.*

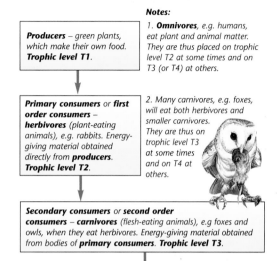

Secondary consumers or **second order consumers** – **carnivores** *(flesh-eating animals), e.g foxes and owls, when they eat herbivores. Energy-giving material obtained from bodies of* **primary consumers**. **Trophic level T3.**

Tertiary consumers or **third order consumers** – *carnivores, e.g. foxes and owls, when they eat smaller carnivores. Energy-giving material is obtained by most indirect method – from bodies of secondary consumers, i.e. animals which ate animals which ate* **producers**. **Trophic level T4.**

Nitrogen cycle

The constant circulation of the element nitrogen through living things, the soil and the atmosphere.

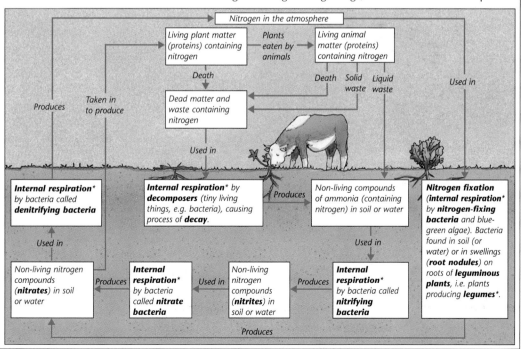

Nitrogen in the atmosphere

Living plant matter (proteins) containing nitrogen — Plants eaten by animals — Living animal matter (proteins) containing nitrogen

Death — Death — Solid waste — Liquid waste — Used in

Produces — Taken in to produce — Dead matter and waste containing nitrogen

Used in

Internal respiration* *by bacteria called* **denitrifying bacteria**

Internal respiration* *by* **decomposers** *(tiny living things, e.g. bacteria), causing process of* **decay**.

Produces — Non-living compounds of ammonia (containing nitrogen) in soil or water

Nitrogen fixation (**internal respiration*** *by* **nitrogen-fixing bacteria** *and blue-green algae). Bacteria found in soil (or water) or in swellings (***root nodules***) on roots of* **leguminous plants**, *i.e. plants producing* **legumes***.

Used in — Used in

Non-living nitrogen compounds (**nitrates**) in soil or water — Produces — **Internal respiration*** *by bacteria called* **nitrate bacteria** — Used in — Non-living nitrogen compounds (**nitrites**) in soil or water — Produces — **Internal respiration*** *by bacteria called* **nitrifying bacteria**

Produces

LIFE AND LIFE CYCLES

All living things show the same basic **characteristics of life**. These are respiration, feeding, growth, sensitivity (irritability), movement, excretion and reproduction. The **life cycle** of a plant or animal is the progression from its formation to its death, with all the changes this entails. (In some cases, these are drastic – see **metamorphosis**, page 49.) Below are some terms used to group plants and animals together according to their life cycle, or to describe characteristics of certain life cycles.

Perennials

Plants which live for many years. **Herbaceous perennials**, e.g. foxgloves, lose all the parts above ground at the end of each growing season, and grow new shoots at the start of the next. **Woody perennials**, e.g. trees, produce new growth (**secondary tissue***) each year from permanent stems.

*Foxgloves are **perennials**.*

Biennials

Plants which live for two years, e.g. carrots. In the first year, they grow and store up food. In the second, they produce flowers and seeds, and then die.

*Carrots are **biennials**.*

Annuals

Plants which live for one year, e.g. lobelias. In this time they grow from seed, produce flowers and seeds, and then die.

*Lobelias are **annuals**.*

Herbaceous

A term describing plants, e.g. phlox, which do not develop **secondary tissue*** above the ground, i.e. they are "like a herb", as distinct from shrubs and trees (**woody perennials**).

*Phlox is a **herbaceous** plant.*

Deciduous

A term describing **perennials** whose leaves lose their **chlorophyll*** and fall off at the end of each growing season, e.g. horse chestnuts.

Horse chestnut

Evergreen

A term describing **perennials** which do not shed their leaves at the end of a growing season, e.g. firs.

Grand fir

Ephemeral

Living for a very short time. Ephemeral plants are found in places which are hot and dry for most of the year (or for many years). The right growing conditions do not exist for long, so they must grow and produce seeds in a very short time. The only truly ephemeral animals are mayflies. Their adult life span is between a few minutes and one day.

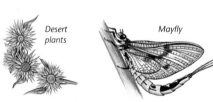

Desert plants

Mayfly

Anadromous

A term describing fish which live in the sea but swim upriver to breed, e.g. salmon. This is a form of **migration**, and the opposite is **catadromous** (going from river to sea).

Salmon swimming upriver

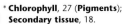

* **Chlorophyll**, 27 (Pigments);
Secondary tissue, 18.

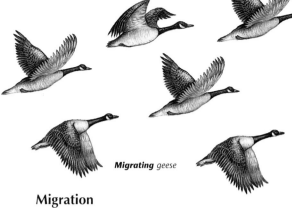

Migrating geese

Migration

Travelling seasonally from one region to another. This normally involves leaving an area in winter to find food elsewhere, and returning in the spring to breed. Migration is part of the life cycle of many animals, especially birds.

Dormancy

A period, or periods, of suspended activity which is a natural part of the life cycle of many plants and animals. Dormancy in plants occurs when conditions are unfavourable for growth (normally in winter). In animals, dormancy usually occurs because of food scarcity, and is either called **hibernation** or **aestivation**. Hibernation is dormancy in the winter (typical of many animals, e.g. some **mammals***), and aestivation is dormancy in drought conditions (occurs mainly in insects).

Dormouse in **hibernation**

Life styles

The world has a vast diversity of living things, each one with its own style of life. This situation is a result of **adaptive radiation**. The living things can be grouped together according to shared characteristics, either by formal classification, which is based mainly on their inferred ancestry (see charts, pages 112-113) or by more informal groupings, based on general life styles (see list, page 114).

Adaptive radiation or evolutionary adaptation

The gradual process which has produced many different forms of living thing from one prehistoric starting point. Each has become **specialized**, i.e. has evolved the best form to cope with its environment, e.g. streamlined shapes for swimming and flying.

Birds' wings are shaped for flight.

The salmon's streamlined body helps it to swim efficiently.

Protective adaptations

Protective measures, such as thorns or poison stings, developed by many animals. All adaptations become established in successive generations because those creatures with them are the most likely to survive long enough to breed (and perpetuate the adaptation). This is the basis of Darwin's theory of **natural selection** (also called **Darwinism**), first expounded in the mid-nineteenth century.

Thorns on rose stalks

A bee's sting protects it from predators.

Mimicry

A special type of adaptation, in which a plant or animal (the **mimic**) has developed a resemblance to another plant or animal (the **model**). This is used especially for protection (e.g. many unprotected insects have adopted the colouring of those with stings), but also for other reasons (bee orchids are mimics for reproduction purposes – see page 31).

Model

Wasp (protected with sting)

Mimic

Hoverfly (unprotected)

***Mammals**, 113.

THE STRUCTURE OF LIVING THINGS

A living thing capable of a separate existence is called an **organism**. All organisms are made up of **cells** – the basic units of life, which carry out all the vital chemical processes. The simplest organisms have just one cell (they are **unicellular** or **acellular**), but very complex ones, e.g. humans, have many billions. They are **multicellular** and their cells are of many different types, each type specially adapted for its own particular job. Groups of cells of the same type (together with non-living material) make up the different **tissues** of the organism, e.g. muscle tissue. Several different types of tissue together form an **organ**, e.g. a stomach, and a number of organs together form a **system**, e.g. a digestive system.

*Protozoa and some algae are **unicellular** organisms.*

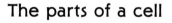

The parts of a cell

All cells are made up of the same basic parts: the **cell membrane**, **cytoplasm** and **nucleus**. Each of these parts has a specific role to play.

Cell membrane
Also called the **plasma membrane** or **plasmalemma**. The outer skin of a cell. It is **semipermeable***, i.e. selective about which substances it allows through.

Cytoplasm
The material where all the chemical reactions vital to life occur (see **organelles**). It generally has a gel-like outer layer and a more liquid inner one (see **ectoplasm** and **endoplasm** – pictures, page 40).

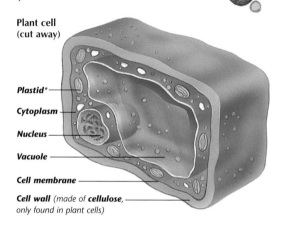

Plant cell (cut away)

Plastid*
Cytoplasm
Nucleus
Vacuole
Cell membrane
Cell wall (made of **cellulose**, only found in plant cells)

Animal cell (cut away)

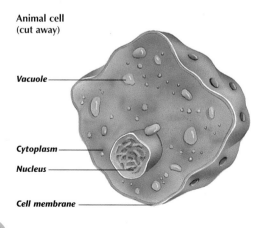

Vacuole
Cytoplasm
Nucleus
Cell membrane

Nucleus (pl. nuclei)
The cell's control centre. Its double-layered outer skin (**nuclear membrane**) encloses a jelly-like fluid (**nucleoplasm** or **karyolymph**), which contains one or more **nucleoli*** and the genetic material **DNA***. This is held in **chromosomes*** – bodies which form a thread-like mass called **chromatin** when the cell is not dividing.

Vacuoles
Fluid-filled sacs in the **cytoplasm**. They are small and temporary in animal cells, and either transfer substances (see **Golgi complex**) or contain fluid brought in (see **pinocytosis**, page 101). Most plant cells have one large, permanent vacuole, filled with **cell sap** (dissolved minerals and sugars).

* **Chromosomes**, 96; **DNA**, 96 (**Nucleic acids**); **Nucleoli**, **Plastids**, 12; **Semipermeable**, 101 (**Diffusion**).

Organelles

The **organelles** are tiny bodies in the **cytoplasm**. Each type (listed below and on page 12) has a vital role to play in the chemical reactions within the cell.

Lysosomes

Round sacs containing powerful **enzymes***. They take in foreign bodies, e.g. bacteria, to be destroyed by the enzymes. Their outer skins do not usually let the enzymes out into the cell (to break down its contents), but if the cell becomes damaged, the skins disappear and the cell digests itself.

Ribosomes

Tiny round particles (most are attached to the **endoplasmic reticulum**). They are involved in building up proteins from amino acids (see page 102). Coded information (held by the **DNA** in the **nucleus**) is sent to the ribosomes in strands of a substance called **messenger RNA** (**mRNA**). These pass on the codes so that the ribosomes join the amino acids in the correct order to produce the right proteins. **RNA*** is present in at least two other forms in the cells. The ribosomes are made of **ribosomal RNA** (see **nucleoli***), and molecules of **transfer RNA** (**tRNA**) carry the amino acids to the ribosomes.

Endoplasmic reticulum or ER

A complex system of flat sacs, joining up with the **nuclear membrane** (see **nucleus**). It provides a large surface area for **enzyme*** reactions. **Rough ER** has **ribosomes** on the surface, where amino acids are combined to make proteins (see page 102). ER with no ribosomes is **smooth ER**. Smooth ER bears enzymes for many other cell processes.

Golgi complex

Also called a **Golgi apparatus**, **Golgi body** or **dictyosome**. A special set of membranous sacs, which collects, modifies and distributes the substances made by the ER (e.g. proteins). The substances fill the sacs, which gradually swell up at their outside edges until pieces "pinch off". These pieces, called **vesicles**, then travel out of the cell via the **cytoplasm** and **cell membrane**.

Golgi complex   Centriole   Cell membrane   Lysosome

Animal cell showing the organelles in the cytoplasm (diagram is not to scale)

Nucleus (double membrane cut away). The **nucleoplasm** and **chromosomes** are not shown.

Nucleolus

Mitochondrion (cross section)

Ribosomes

Endoplasmic reticulum (rough ER)

Vacuole

Endoplasmic reticulum (smooth ER)

* **Enzymes**, 105; **Nucleoli**, 12; **RNA**, 96 (**Nucleic acids**).

Organelles (continued)

Centrioles

Two bodies in animal and primitive plant cells which are vital to **cell division** (see right). In animal cells, they lie just outside the **nucleus***. Each lies in a dense area of **cytoplasm*** (**centrosome**) and is made up of two tiny cylinders, forming a +-shape or T-shape. Each cylinder is made up of nine sets of three tiny tubes (**microtubules**).

Centriole

Nucleoli (sing. nucleolus)

One or more small, round bodies in the **nucleus***. They produce the component parts of the **ribosomes*** (made of **ribosomal RNA**), which are then transported out of the nucleus and assembled in the **cytoplasm***.

Nucleolus

Mitochondria (sing. mitochondrion)

Rod-shaped bodies with a double layer of outer skin. The inner layer forms a series of folds (**cristae**, sing. **crista**), providing a large surface area for the vital chemical reactions which go on inside the mitochondria (called the cell's "powerhouses"). They are the places where simple substances taken into the cell are broken down to provide energy. For more about this, see **aerobic respiration**, page 106.

Cristae

Mitochondrion

Plastids

Tiny bodies in plant cell **cytoplasm***. Some (**leucoplasts**) store starch, oil or proteins. Others – **chloroplasts*** – contain **chlorophyll*** (used in making food).

Plastid (chloroplast)*

Cell division

Cell division is the splitting up of one cell (the **parent cell**) into two identical **daughter cells**. There are two types of cell division, both involving the division of the **nucleus*** (**karyokinesis**) followed by the division of the **cytoplasm*** (**cytokinesis**). The first type of cell division (**mitosis**) is described on these two pages. It produces new cells for growth and also to replace the millions of cells which die each day (from damage, disease or simply because they are "worn out"). It is also the means of **asexual reproduction*** in many single-celled organisms. The second, special type of cell division produces the **gametes*** (sex cells) which will come together to form a new living thing. For more about this, see pages 94-95.

Mitosis

The division of the **nucleus*** when a plant or animal cell divides for growth or repair. It ensures that the two new nuclei (**daughter nuclei**) are each given the same number of **chromosomes*** (the bodies which carry the "coded" hereditary information). Each receives the same number of chromosomes as were in the original nucleus, called the **diploid number**. Every living thing has its own characteristic diploid number, i.e. all its cells (with the exception of the **gametes***) contain the same, specific number of chromosomes, grouped in identical pairs called **homologous chromosomes**. Humans have 46 chromosomes, in 23 pairs. Although mitosis is a continuous process, it can be divided for convenience sake into four phases. Before mitosis, however, there is always an **interphase**.

Interphase

The periods between cell divisions. Interphases are very active periods, during which the cells are not only carrying out all the processes needed for life, but are also preparing material to make "copies" of all their components (so both new cells formed after division will have all they need). Just before **mitosis** begins, the **chromatin*** threads in the **nucleus*** also duplicate, so that, after coiling up, each **chromosome*** will consist of two **chromatids** (see **prophase**). The **centriole** duplicates itself during interphase.

Phases of mitosis

1. Prophase

The **nuclear membrane*** begins to break down and the threads of **chromatin*** in the **nucleus*** coil up to form **chromosomes***. Each has already duplicated to form two identical, long coils (**chromatids**), joined by a sphere (**centromere**). The two **centrioles** move to opposite poles (ends) of the cell, as **spindle microtubules** form between them.

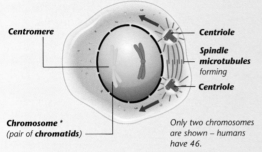

Centromere

Centriole

Spindle microtubules forming

Centriole

Chromosome * (pair of **chromatids**)

Only two chromosomes are shown – humans have 46.

2. Metaphase

The **nuclear membrane*** disappears and the **spindle microtubules** surround the **chromosomes*** (paired **chromatids**). The chromosomes move towards its equator and become attached by their **centromeres** to the **spindle microtubules**.

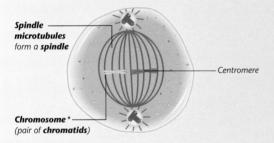

Spindle microtubules form a **spindle**

Centromere

Chromosome * (pair of **chromatids**)

3. Anaphase

The **centromeres** split and the two **chromatids** from each pair (now called **daughter chromosomes**) move to opposite poles of the **spindle**, seemingly "dragged" there by the contracting **spindle microtubules**.

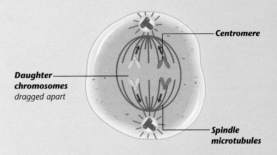

Centromere

Daughter chromosomes *dragged apart*

Spindle microtubules

4. Telophase

The **spindle microtubules** disappear and a new **nuclear membrane*** forms around each group of **daughter chromosomes**. This creates two new **nuclei*** (**daughter nuclei**), inside which the chromosomes uncoil and once again form a thread-like mass (**chromatin***).

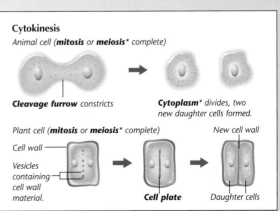

Cleavage furrow *(see cytokinesis, below)*

New **nuclei**

Daughter chromosomes *(before uncoiling)*

Cytokinesis

The division of the **cytoplasm*** of a cell, which forms two new cells around the new **nuclei*** created by **mitosis** (or **meiosis***). In animal cells, a **cleavage furrow** forms around the cell's equator and then constricts as a ring until it cuts completely through the cell. In plant cells, a dividing line called the **cell plate** forms down the centre of the cell, and a new **cell wall*** is built up along each side of it.

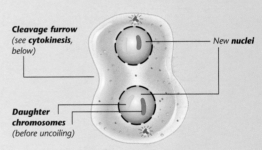

Cytokinesis

*Animal cell (**mitosis** or **meiosis*** complete)*

Cleavage furrow *constricts*

Cytoplasm* *divides, two new daughter cells formed.*

*Plant cell (**mitosis** or **meiosis*** complete)*

Cell wall

Vesicles containing cell wall material.

New cell wall

Cell plate

Daughter cells

* **Cell wall**, 10; **Chromatin**, 10 (**Nucleus**); **Chromosomes**, 96;
Cytoplasm, 10; **Meiosis**, 94; **Nuclear membrane**, 10 (**Nucleus**).

VASCULAR PLANTS

With the exception of simple plants such as algae, fungi, mosses and liverworts (see classification chart, page 112), all plants are **vascular plants**. That is, they all have a complex system of special fluid-carrying tissue called **vascular tissue**. For more about how the fluids travel within the vascular tissue, see pages 24-25.

Vascular tissue

Special tissue which runs throughout a **vascular plant**, carrying fluids and helping to support the plant. In young stems, it is normally arranged in separate units called **vascular bundles**; in older stems, these join up to form a central core (**vascular cylinder***). In young roots, the arrangement of the tissue is slightly different, but a central core is also formed later. For more about the vascular tissue in older plants, see page 18. The vascular tissue is of two different types – **xylem** and **phloem**. They are separated by a layer of tissue called the **cambium**.

*Tulips are **monocotyledons***. Their **vascular bundles** are irregularly arranged within the stem. In **dicotyledons***, by contrast, the bundles are more regular (see root and stem sections, right).*

Cross section of young stem, or young part of a stem (dicotyledon*)

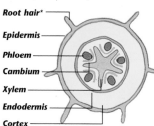

Cortex
Vascular bundle
Xylem
Phloem
Cambium

Young stem, or young part of a stem (dicotyledon*)

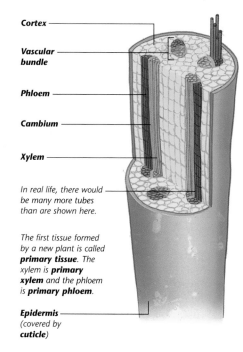

Cortex
Vascular bundle
Phloem
Cambium
Xylem

In real life, there would be many more tubes than are shown here.

*The first tissue formed by a new plant is called **primary tissue**. The xylem is **primary xylem** and the phloem is **primary phloem**.*

Epidermis *(covered by **cuticle**)*

Cross section of young root, or young part of a root (dicotyledon*)

Root hair*
Epidermis
Phloem
Cambium
Xylem
Endodermis
Cortex

Longitudinal section of young root, or young part of a root (dicotyledon*)

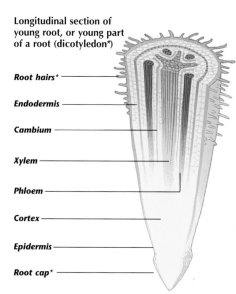

Root hairs*
Endodermis
Cambium
Xylem
Phloem
Cortex
Epidermis
Root cap*

* **Dicotyledon, Monocotyledon**, 33 (**Cotyledon**); **Root cap, Root hairs**, 17; **Vascular cylinder**, 18.

Constituents of vascular tissue

Xylem

A tissue which carries water up through a plant. In flowering plants it is made up of **vessels** or **tracheids**, with long, thin cells (**fibres**) providing support between them. Non-flowering plants have only tracheids. In older stems, the vessels become filled in, forming **heartwood***.

Vessels and tracheids

Tubes in the **xylem** which carry water. Their walls are strengthened with a hard substance called **lignin**. They occur as columns of cells whose contents have died. Vessels are shorter and wider than tracheids.

Cambium

A layer of narrow, thin-walled cells between the **xylem** on the inside and the **phloem** on the outside. The cells are able to divide, making more xylem and phloem. Such an area of cells is called a **meristem***.

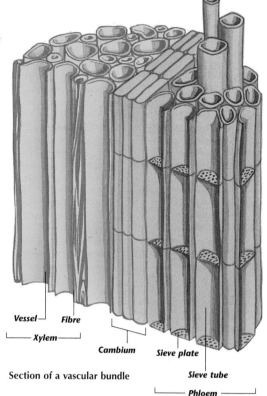

Section of a vascular bundle

Labels: Vessel — Fibre — Xylem — Cambium — Sieve plate — Sieve tube — Phloem

Phloem

A tissue which distributes the food made in the leaves to all parts of the plant. It consists of fluid-carrying **sieve tubes**, each with a **companion cell** beside them, and other cells packed around them for support.

Sieve tubes

Cells in long columns in the **phloem**. They are living cells with **cell walls*** and a thin layer of **cytoplasm** but no **nucleus***. The end walls between the cells, called **sieve plates**, have tiny holes in them to allow substances through.

Other tissues in vascular plants

Epidermis

A thin surface layer of tissue around all parts of a plant. In some areas, especially the leaves, it has many tiny holes, called **stomata***. In older stems, the epidermis is replaced by **phellem***. In older roots, it is replaced by **exodermis*** and then by phellem.

Cortex

A layer of tissue inside the **epidermis** of stems and roots. It consists mainly of **parenchyma**, a type of tissue with large cells and many air spaces. In some plants there is also some **collenchyma**, a type of supporting tissue with long, thick-walled cells. The cortex tends to get compressed and replaced by other tissues as a plant gets older.

Endodermis

The innermost layer of root **cortex**. Fluids which have seeped in between the cortex cells, instead of through them, are directed by its special **passage cells** into the central area of **vascular tissue**.

Pith or medulla

A central area of tissue found in stems, but not usually in roots. It is generally only called pith once the stem has developed a **vascular cylinder***. It is made up of **parenchyma** (see cortex), and is sometimes used to store food.

Cuticle

A thin outer layer of a waxy substance called **cutin** made by the **epidermis** above ground. It prevents too much water from being lost.

---

* **Exodermis**, 17 (**Piliferous layer**); **Heartwood**, 19; **Meristem**, 16; **Nucleus**, 10; **Phellem**, 19; **Stomata**, 21; **Vascular cylinder**, 18.

STEMS AND ROOTS

The **stem** and **roots** of a plant are its main supporting structures, as well as being important in transporting fluids (see pages 14-15 and 24-25). Their various parts are listed here. For more about the development of the stem and roots as plants get older, see pages 18-19.

Meristem
Any area in a plant from which new growth arises. The cells of a meristem are able to divide, producing new cells. A meristem found at the tip of the root (the **growing point**) or the stem (part of a **terminal bud**) is known as an **apical meristem**.

Stem attachments

Shoot
A new stem growing out of a seed or off the main stem of a plant.

Bud
A small outgrowth on a stem. It develops either into a new **shoot** or a flower.

Terminal bud
A **bud** growing at the end of a stem or **shoot**.

Axillary bud
Also called a **lateral bud** or **secondary bud**. A **bud** situated in an **axil** – the angle between a **shoot** or leaf stalk and the stem it is growing from.

Node
A place on a stem where a leaf, with or without a leaf stalk, has been produced.

Internode
The area of a stem or **shoot** between two **nodes**.

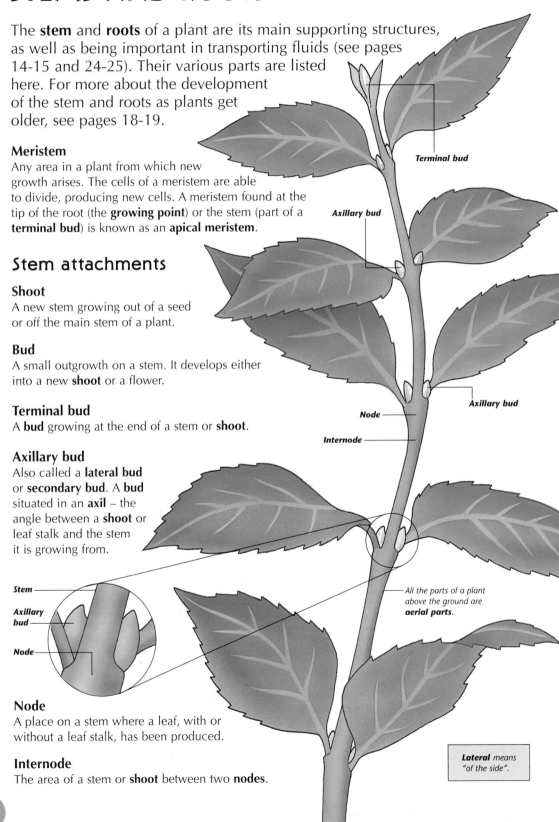

Terminal bud

Axillary bud

Axillary bud

Node

Internode

Stem

Axillary bud

Node

All the parts of a plant above the ground are **aerial parts**.

Lateral *means "of the side".*

Parts of a root

Root cap
A layer of cells which protects the root tip as it is pushed down into the ground.

Growing point
An area just behind a root tip where the cells divide to produce new growth.

Zone of elongation
The area of new cells produced by the **growing point**, and located just behind it. The cells stretch lengthwise as they take in water, since their **cell walls*** are not yet hard. This elongation pushes the root tip further down into the soil.

Piliferous layer
The youngest layer of the **epidermis***, or outer skin, of a root. It is the area which produces **root hairs**. It is found just behind the **zone of elongation**. As the walls of the elongating cells harden, the outermost cells become the piliferous layer. The older piliferous layer (higher up the root) is slowly worn away, to be replaced by a layer of hardened cells called the **exodermis** (the outermost layer of the **cortex***).

Root hairs
Long outgrowths from the cells of the **piliferous layer**. They take in water and minerals.

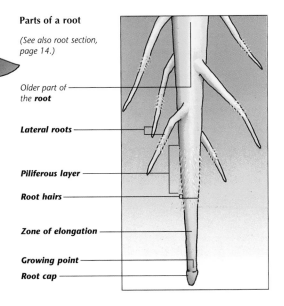

Parts of a root

(See also root section, page 14.)

Older part of the **root**

Lateral roots

Piliferous layer

Root hairs

Zone of elongation

Growing point

Root cap

Types of root

Tap root
A first root, or **primary root**, which is larger than the small roots, called **lateral roots** or **secondary roots**, which grow out of it. Many vegetables are swollen tap roots.

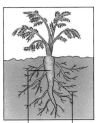

Tap root (carrot)   **Lateral root**

Fibrous roots
A system made up of a large number of equal-sized roots, all producing smaller **lateral roots**. The first root is not prominent, as it is in a **tap root** system.

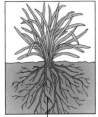

Fibrous roots

Adventitious roots
Roots which grow directly from a stem. Adventitious roots grow out of **bulbs*** (which are special types of stem), or from gardeners' cuttings.

Adventitious roots

Aerial roots
Roots which grow from stems and do not normally grow into the ground. They can be used for climbing, e.g. in an ivy. Many absorb moisture from the air.

Ivy   **Aerial roots**

Prop roots
Special types of **aerial root**. They grow out from a stem and then down into the ground, which may be under water. Prop roots support a heavy plant, e.g. a mangrove.

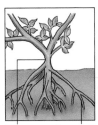

Mangrove   **Prop roots**

* **Bulb**, 35; **Cell wall**, 10; **Cortex**, **Epidermis**, 15.

INSIDE AN OLDER PLANT

A plant which lives for many years, such as a tree, forms **secondary tissue** as it grows. This consists of new layers of tissue to supplement the original tissue, or **primary tissue***. New supportive and fluid-carrying **vascular tissue*** is formed towards the centre of the plant and new protective tissue is produced around the outside. The production of the new vascular tissue is called **secondary thickening**, and results in what is known as a **woody plant**.

New central tissue

Vascular cylinder

A vascular cylinder develops as the first step of **secondary thickening** in stems. More **cambium*** forms between the **vascular bundles***, and this then gives rise to more **xylem*** and **phloem***, forming a continuous cylinder.

Secondary thickening

The year-by-year production of more fluid-carrying **vascular tissue*** in plants which live for many years, resulting in a gradual increase in the diameter of the stem and roots. Each year, new layers of **xylem*** (**secondary xylem**) and **phloem*** (**secondary phloem**) are produced by the dividing cells of the **cambium*** between them. This process differs slightly between stems and roots, but the result throughout the plant is an ever-enlarging core of vascular tissue (which slowly "squeezes out" the **pith*** in stems). Most of this core is xylem, now also known as **wood**. The area of phloem does not widen much at all, because the xylem pushing outwards wears it away.

Annual rings

The concentric circles which can be seen in a cross section of an older plant. Each ring is one year's new growth of **xylem***, and has two separate areas – **spring wood** and **summer wood**. Soft **spring wood** (or **early wood**) forms rapidly early in the growing season and has widely-spaced cells. Harder **summer wood** (or **late wood**) is produced later on. Its cells are more densely packed.

The roots and trunk of this tree thicken as it grows.

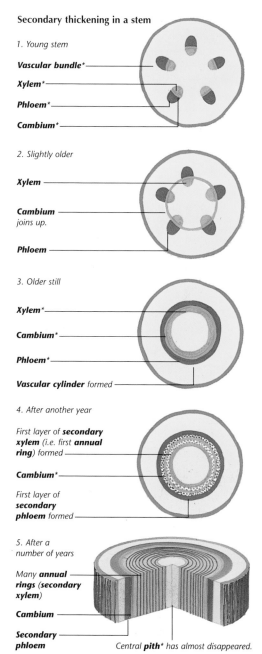

Secondary thickening in a stem

1. Young stem

Vascular bundle*
Xylem*
Phloem*
Cambium*

2. Slightly older

Xylem
Cambium joins up.
Phloem

3. Older still

Xylem*
Cambium*
Phloem*
Vascular cylinder formed

4. After another year

First layer of **secondary xylem** (i.e. first **annual ring**) formed
Cambium*
First layer of **secondary phloem** formed

5. After a number of years

Many **annual rings** (secondary xylem)
Cambium
Secondary phloem
*Central **pith*** has almost disappeared.*

---

* **Cambium, Phloem, Pith**, 15; **Primary tissue**, 14;
Vascular bundles, 14 (**Vascular tissue**); **Xylem**, 15.

New outer tissue

As well as new **vascular tissue***, an older plant also forms extra areas of tissue around its outside to help protect it. These are called **phelloderm**, **phellogen** and **phellem** respectively (working from the inside). The three areas together are known as the **periderm**.

Phellogen or cork cambium

A cell layer which arises towards the outside of the stem and roots of older plants. It is a **meristem***, i.e. an area of cells which keep on dividing. It produces two new layers – **phelloderm** and **phellem**.

Phelloderm

A new cell layer produced by **phellogen** on its inside. It supplements the **cortex*** and is sometimes called **secondary cortex**.

Phellem or cork

A new cell layer produced by **phellogen** on its outside. The cells undergo **suberization**, i.e. become impregnated with a waxy substance called **suberin**. This makes the outer layer waterproof. The phellem cells slowly die and replace the previous outer cell layer (**epidermis*** in stems and **exodermis*** in roots). Dead phellem cells are called **bark**.

Bark stops the tree from drying out and protects it from disease. It cannot grow or stretch, so it splits or peels as the trunk gets wider, and new bark grows underneath.

Silver birch bark

English oak bark

Scots pine bark

Beech bark

Tree (many years old)

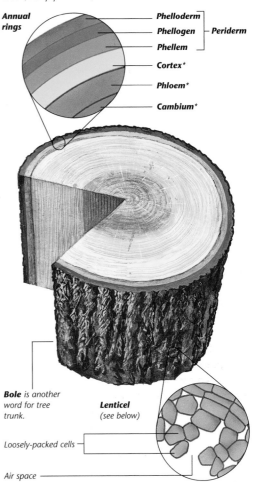

Annual rings — Phelloderm, Phellogen, Phellem — Periderm; Cortex*; Phloem*; Cambium*

Bole is another word for tree trunk.

Lenticel (see below)

Loosely-packed cells

Air space

Lenticels

Tiny raised openings in the **phellem** through which an older plant exchanges oxygen and carbon dioxide. Inside them, a channel of loosely-packed cells allows the gases to move across the outer tissues to or from the **cortex***, which also has air spaces.

Types of wood

Heartwood

The oldest, central part of the **xylem*** in an older plant. The **vessels*** are filled in and no longer carry fluids, but they still provide support.

Heartwood

Sapwood

The outer area of **xylem*** in an older plant, whose **vessels*** still carry fluids. Sapwood also supports the tree and holds the tree's food reserves.

Sapwood

---

* **Cambium, Cortex, Epidermis**, 15; **Exodermis**, 17 (**Piliferous layer**); **Phloem**, 15;
Meristem, 16; **Vascular tissue**, 14; **Vessels, Xylem**, 15.

LEAVES

The **leaves** of a plant, collectively known as its **foliage**, are specially adapted to manufacture food. They do this by a process called **photosynthesis**. For more about this, see pages 26-27. There are many different shapes and sizes of leaves, but only two different types. **Simple leaves** consist of a single leaf blade, or **lamina**, and **compound leaves** are made up of small leaf blades called **leaflets**, all growing from the same leaf stalk. You can find out more about some of the different leaf shapes on page 22.

Simple leaf (holly)

Compound leaf (horse chestnut)

Inside a leaf

Veins

Long strips of **vascular tissue*** inside a leaf (see picture, right), supplying it with water and minerals and removing the food made inside it. Some leaves have long, parallel veins, e.g. those of grasses, but most have a central vein inside a **midrib** (an extension of the leaf stalk), with many smaller branching veins.

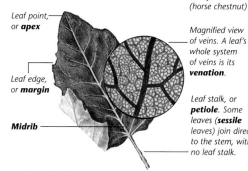

Leaf point, or *apex*

Leaf edge, or *margin*

Magnified view of veins. A leaf's whole system of veins is its **venation**.

Midrib

Leaf stalk, or **petiole**. *Some leaves (*sessile* leaves) join directly to the stem, with no leaf stalk.*

Spongy layer

A layer of irregular-shaped **spongy cells** and air spaces where gases circulate. The **spongy** and **palisade layers** together are the **mesophyll**.

Palisade layer

A cell layer just below the upper surface of a leaf. It is made up of regular, oblong-shaped **palisade cells**. These contain many **chloroplasts***.

Palisade cell

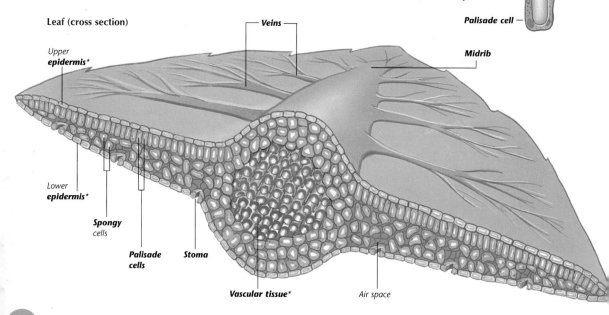

Leaf (cross section)

Upper **epidermis***

Veins

Midrib

Lower **epidermis***

Spongy *cells*

Palisade *cells*

Stoma

Vascular tissue*

Air space

Chloroplasts, 26; **Epidermis**, 15; **Vascular tissue**, 14.

Stomata (sing. stoma)

Tiny openings in the **epidermis*** (outer skin), through which the exchange of water (**transpiration***) and gases takes place. Stomata are mainly found on the underside of leaves.

Guard cells

Pairs of crescent-shaped cells. The members of each pair are found on either side of a **stoma**, which they open and close by changing shape. This controls water and gas exchange. Guard cells are the only surface cells with **chloroplasts***.

Leaf trace

An area of **vascular tissue*** which branches off that of a stem to become the central **vein** of a leaf.

Abscission layer

A layer of cells at the base of a leaf stalk which separates from the rest of the plant at a certain time of year (stimulated by a **hormone*** called **abscisic acid**). This makes the leaf fall off, forming a **leaf scar** on the stem.

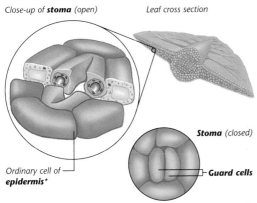

Close-up of stoma (open)   *Leaf cross section*

Stoma (closed)

Ordinary cell of epidermis*   *Guard cells*

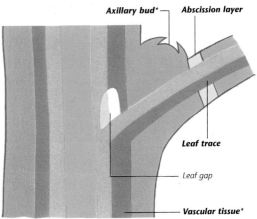

Axillary bud*   *Abscission layer*

Leaf trace

Leaf gap

Vascular tissue*

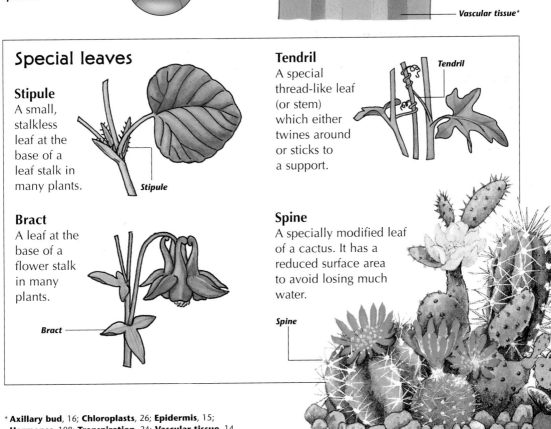

Special leaves

Stipule
A small, stalkless leaf at the base of a leaf stalk in many plants.

Stipule

Bract
A leaf at the base of a flower stalk in many plants.

Bract

Tendril
A special thread-like leaf (or stem) which either twines around or sticks to a support.

Tendril

Spine
A specially modified leaf of a cactus. It has a reduced surface area to avoid losing much water.

Spine

* **Axillary bud**, 16; **Chloroplasts**, 26; **Epidermis**, 15; **Hormones**, 108; **Transpiration**, 24; **Vascular tissue**, 14.

Types of compound leaf

Shown on this page are some types of **compound leaf** (leaves made up of **leaflets***), as well as some common leaf arrangements and leaf edges, or **margins**. The pictures are not to scale.

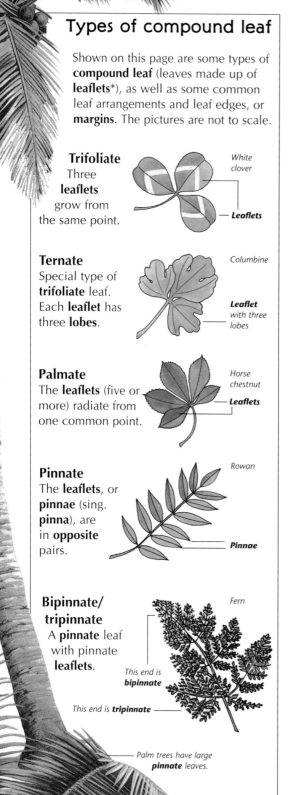

Trifoliate
Three leaflets grow from the same point.

White clover

Leaflets

Ternate
Special type of **trifoliate** leaf. Each **leaflet** has three **lobes**.

Columbine

Leaflet with three lobes

Palmate
The **leaflets** (five or more) radiate from one common point.

Horse chestnut

Leaflets

Pinnate
The **leaflets**, or **pinnae** (sing. **pinna**), are in **opposite** pairs.

Rowan

Pinnae

Bipinnate/ tripinnate
A **pinnate** leaf with pinnate **leaflets**.

Fern

This end is **bipinnate**

This end is **tripinnate**

Palm trees have large **pinnate** leaves.

Leaf arrangements

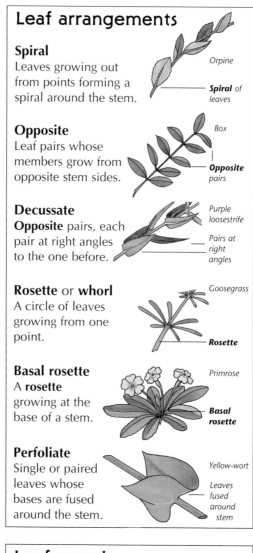

Spiral
Leaves growing out from points forming a spiral around the stem.

Orpine

Spiral of leaves

Opposite
Leaf pairs whose members grow from opposite stem sides.

Box

Opposite pairs

Decussate
Opposite pairs, each pair at right angles to the one before.

Purple loosestrife

Pairs at right angles

Rosette or whorl
A circle of leaves growing from one point.

Goosegrass

Rosette

Basal rosette
A **rosette** growing at the base of a stem.

Primrose

Basal rosette

Perfoliate
Single or paired leaves whose bases are fused around the stem.

Yellow-wort

Leaves fused around stem

Leaf margins

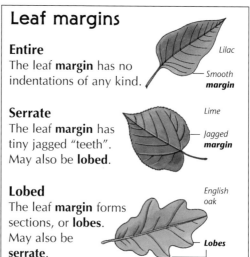

Entire
The leaf **margin** has no indentations of any kind.

Lilac

Smooth **margin**

Serrate
The leaf **margin** has tiny jagged "teeth". May also be **lobed**.

Lime

Jagged **margin**

Lobed
The leaf **margin** forms sections, or **lobes**. May also be **serrate**.

English oak

Lobes

* **Leaflets**, 20.

PLANT SENSITIVITY

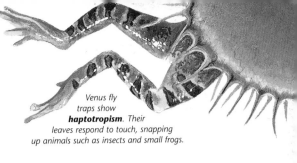

Plants have no nervous system, but they do still show **sensitivity**, i.e. they react to certain forms of stimulation. They do this by moving specific parts or by growing. This is called **tropism**. **Positive tropism** is movement or growth towards the stimulus and **negative tropism** is movement or growth away from it.

Venus fly traps show **haptotropism**. *Their leaves respond to touch, snapping up animals such as insects and small frogs.*

Hydrotropism
Response to water. For example, some roots may grow out sideways if there is more water in that direction.

Geotropism
Response to the pull of gravity. This is shown by all roots, i.e. they all grow down through the soil.

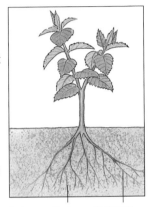

Roots grow down in response to gravity.

Roots grow towards water.

Phototropism
Response to light. When the light is sunlight, the response is called **heliotropism**. Most leaves and stems show this by curving around to grow towards the light.

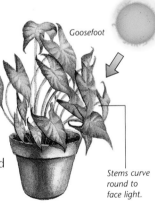

Goosefoot

Stems curve round to face light.

Haptotropism or thigmotropism
Response to touch or contact. For example, the sticky hairs of a sundew plant curl around an insect when it comes into contact with them.

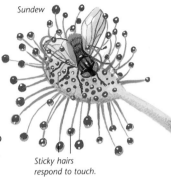

Sundew

Sticky hairs respond to touch.

Photoperiodism
The response of plants to the length of day or night (**photoperiods**), especially with regard to the production of flowers. It depends on a number of things, e.g. the plant's age and the temperature of its environment. **Long-night plants** only produce flowers if the night is longer than a certain length (called its **critical length**), **short-night plants** only if it is shorter. It is thought that a "message" to produce flowers is carried to the relevant area by a **hormone***, produced in the leaves when the conditions are right. This hormone has been called **florigen**. Some plants are **night-neutral plants**, i.e. their flowering does not depend on the length of night (see pictures, below).

These three plants each produce flowers according to different **photoperiods***.*

Chrysanthemum (**long-night plant**)

Larkspur (**short-night plant**)

Snapdragon (**night-neutral plant**)

Growth hormones or growth regulators
Substances which promote and regulate plant growth. They are produced in **meristems*** (areas where cells are constantly dividing). **Auxins**, **cytokinins** and **gibberellins** are types of growth hormone.

***Hormones**, 108; **Meristem**, 16.

PLANT FLUID TRANSPORTATION

The transportation of fluids in a plant is called **translocation**. The fluids travel within the **vascular tissue***, made up of **xylem*** and **phloem***. The xylem carries water (with dissolved minerals) from the roots to the leaves. The phloem carries food from the leaves to areas where it is needed.

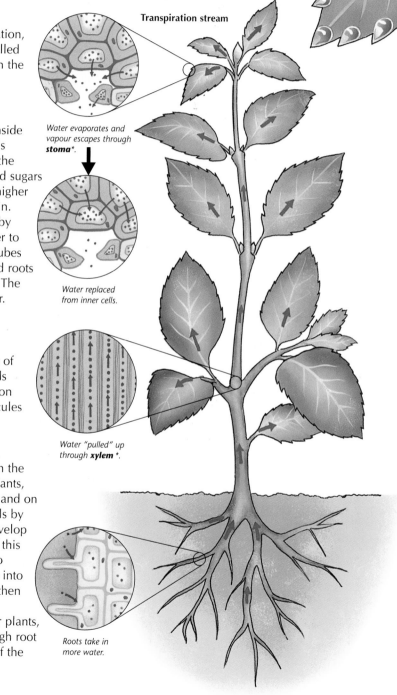

Transpiration

The loss of water by evaporation, mainly through tiny holes called **stomata*** which are found on the undersides of leaves.

Transpiration stream

Water evaporates and vapour escapes through **stoma***.

Transpiration stream

A constant chain of events inside a plant. As the outer leaf cells lose water by **transpiration**, the concentration of minerals and sugars in their **vacuoles*** becomes higher than that of the cells further in. Water then passes outwards by **osmosis***, causing more water to be "pulled" up through the tubes of the **xylem*** in the stem and roots (helped by **capillary action**). The roots then take in more water.

Water replaced from inner cells.

Capillary action

The way that fluids travel up narrow tubes. The molecules of the fluid are "pulled" upwards by the strong force of attraction between them and the molecules of the tube.

Water "pulled" up through **xylem** *.

Root pressure

A pressure which builds up in the roots of some plants. In all plants, water travels in from the soil and on through the layers of root cells by **osmosis***. In plants which develop root pressure, the pressure of this water movement is enough to force the water some way up into the tubes of the **xylem***. It is then "pulled" on upwards by the **transpiration stream**. In other plants, the movement of water through root cells is all due to the "pull" of the transpiration stream.

Roots take in more water.

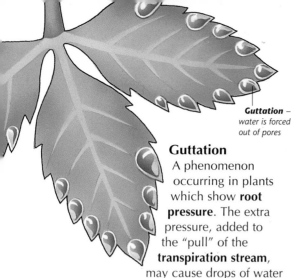

Guttation –
water is forced
out of pores

Guttation

A phenomenon occurring in plants which show **root pressure**. The extra pressure, added to the "pull" of the **transpiration stream**, may cause drops of water to be forced out of water-secreting areas of cells (**hydathodes**) via tiny pores at the tips or along the edges of the leaves.

Turgor

Healthy plant

The state of the cells in a healthy plant when its cells can take in no more water. Each cell is then said to be **turgid**. This means that water has passed by **osmosis*** into the **cell sap*** (dissolved minerals and sugars) in the cell's large central **vacuole***, and the vacuole has pushed as far out as it can go. The vacuole can push out no further because its outward pressure (called **turgor pressure**) is equalled by the opposing force of the rigid **cell wall*** (**wall pressure**). Turgid cells are important because they enable a plant to stand firm and upright.

Turgor

Root cells

Vacuole*
containing
cell sap*

Turgor pressure

Root
hair*

No more water
can enter.

Wall pressure

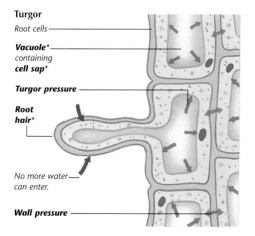

Wilting

Wilting plant

A state of drooping, found in a plant subjected to certain conditions, such as excess heat. The plant is losing more water (by **transpiration**) than it can take in, and the **turgor pressure** (see **turgor**) of its cell **vacuoles*** drops. The cells become limp and can no longer support the plant, so it will droop.

Wilting

Root cells

Reduced **turgor
pressure**

Vacuole* shrinks

Not enough
water coming in

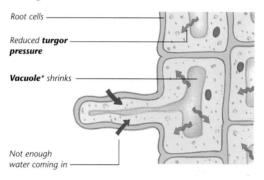

Plasmolysis

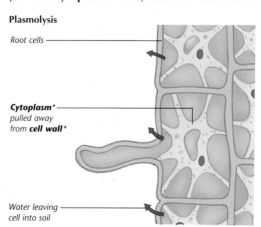

Dying plant

An extreme state in a plant, which may cause it to die. Such a plant is losing a large amount of water, often not only by **transpiration** in excess heat (see **wilting**), but also by **osmosis*** into very dry soil or soil with a very high concentration of minerals. The **vacuoles*** of the plant cells then shrink so much that they pull the **cytoplasm*** away from the **cell walls***.

Plasmolysis

Root cells

Cytoplasm*
pulled away
from **cell wall***

Water leaving
cell into soil

* **Cell sap**, 10 (**Vacuoles**); **Cell wall**, **Cytoplasm**, 10;
 Osmosis, 101; **Root hairs**, 17.

PLANT FOOD PRODUCTION

*Plants need water and carbon dioxide for **photosynthesis**. Lianas have extremely long, twisting stems which carry water to their leaves, where photosynthesis takes place.*

Most plants have the ability to make the food they need for growth and energy (unlike animals, which must take it in). The manufacturing process by which they make their complex food substances from other, simpler substances is called **photosynthesis**.

Photosynthesis
The series of chemical reactions by which green plants make their food. It occurs mainly in the **palisade cells***. Carbon dioxide is combined with water (containing minerals – see below), using energy taken in from sunlight by **chloroplasts**. This produces oxygen as well as the plant's food (see diagram, page 27).

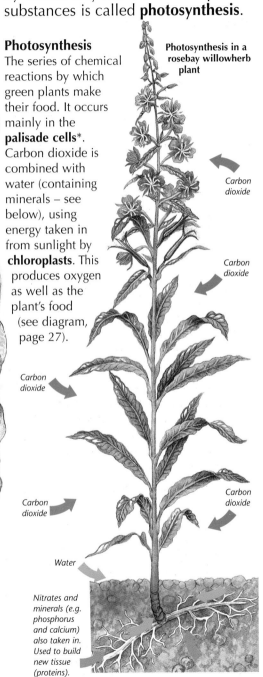

Photosynthesis in a rosebay willowherb plant

Carbon dioxide

Carbon dioxide

Carbon dioxide

Carbon dioxide

Carbon dioxide

Water

Nitrates and minerals (e.g. phosphorus and calcium) also taken in. Used to build new tissue (proteins).

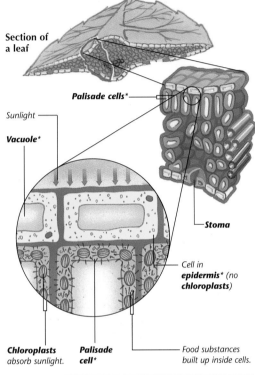

Section of a leaf

Palisade cells*

Sunlight

Vacuole*

Stoma

*Cell in **epidermis*** (no **chloroplasts**)

Chloroplasts absorb sunlight.

Palisade cell*

Food substances built up inside cells.

Chloroplasts
Tiny bodies in plant cells (mainly in the leaves) which contain a green **pigment** called **chlorophyll**. This absorbs the Sun's light energy and uses it to "power" **photosynthesis**. Chloroplasts can move around inside a cell, according to light intensity and direction. See also page 12.

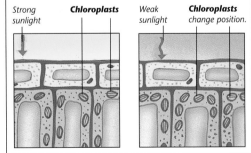

Strong sunlight

Chloroplasts

Weak sunlight

Chloroplasts change position.

*Epidermis, 15; Internal respiration, 106; Palisade cells, 20 (Palisade layer); Vacuoles, 10.

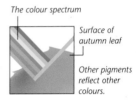

Products of photosynthesis

The process of photosynthesis works in co-ordination with that of **internal respiration***, the breakdown of food for energy. Photosynthesis produces oxygen and carbohydrates (needed for internal respiration), and internal respiration produces carbon dioxide and water (needed for photosynthesis).

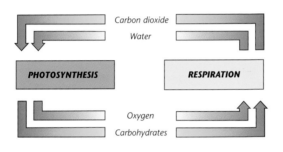

At most times, one of the two processes is occurring at a faster rate than the other. This means that excess amounts of its products are being produced, and not enough of the substances it needs are being made in the plant. In this case, extra amounts must be taken in and excess amounts given off or stored (see pictures 2 and 4 below).

Pigments

Substances which absorb light. White light is actually made up of a spectrum of many different colours. Each pigment absorbs some colours and reflects others.

Chlorophyll is a pigment found in all leaves. It absorbs blue, violet and red light, and reflects green light. This is why leaves look green.

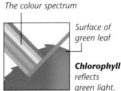

The colour spectrum

Surface of green leaf

Chlorophyll reflects green light.

The colour spectrum

Surface of autumn leaf

Other pigments reflect other colours.

Other pigments, such as **xanthophyll**, **carotene** and **tannin** are also present in leaves. They reflect orange, yellow and red light, but are masked by chlorophyll during the growing season. In autumn, the chlorophyll breaks down, and so the autumn colours appear. Plant pigments are used to give colour to many things, e.g. paints and plastics.

Autumn colours appear when *chlorophyll* breaks down.

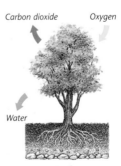

Compensation points

Two points in a 24-hour period (normally around dawn and around dusk) when the processes of **photosynthesis** and **internal respiration*** (see above) are exactly balanced.

Photosynthesis is producing just the right amounts of carbohydrates and oxygen for internal respiration, and this is producing just the right amounts of carbon dioxide and water for photosynthesis.

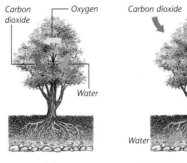
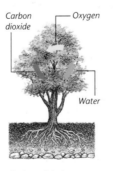

Carbon dioxide — Oxygen

Water

1. Around dawn
(**compensation point**)

Carbon dioxide — Oxygen

Water

2. Midday (bright light, so faster **photosynthesis**)

Carbon dioxide — Oxygen

Water

3. Around dusk
(**compensation point**)

Carbon dioxide — Oxygen

Water

4. Midnight (no light, so no **photosynthesis**)

****Internal respiration**, 106.

FLOWERS

The **flowers** of a plant contain its organs of **reproduction*** (producing new life). In **hermaphrodite** plants, e.g. buttercups and poppies, each flower has both male and female organs. **Monoecious** plants, e.g. maize, have two types of flower on one plant – **staminate** flowers, which have just male organs, and **pistillate** flowers, which have just female organs. **Dioecious** plants, e.g. holly, have staminate flowers on one plant and pistillate flowers on a separate plant.

Receptacle
The expanded tip of the flower stalk, or **peduncle**, from which the flower grows.

Petals
The delicate, usually brightly coloured structures around the reproductive organs. They are often scented (to attract insects) and are known collectively as the **corolla**.

Sepals
The small, leaf-like structures around a bud, known collectively as the **calyx**. In some flowers, e.g. buttercups, they remain as a ring around the opened **petals**; in others, e.g. poppies, they wither and fall off.

Nectaries
Areas of cells at the base of the **petals** which produce a sugary liquid called **nectar**. This attracts insects needed for **pollination***. It is thought that the dark lines down many petals are there to direct an insect to the nectar, and they are called **honey guides**.

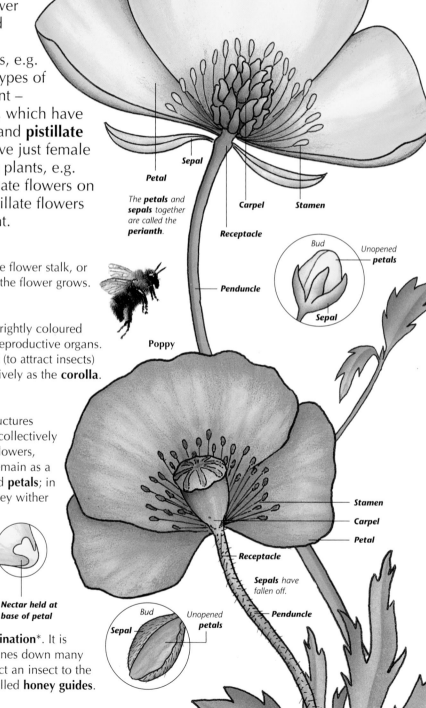

Buttercup

Petal

Sepal

*The **petals** and **sepals** together are called the **perianth**.*

Carpel

Stamen

Receptacle

Penduncle

Bud

Unopened petals

Sepal

Poppy

Nectar held at base of petal

Stamen

Carpel

Petal

Receptacle

Sepals have fallen off.

Bud

Unopened petals

Sepal

Penduncle

* **Pollination, Reproduction**, 30.

The female organs

Carpel or pistil
A female reproductive organ, consisting of an **ovary**, **stigma** and **style**. Some flowers have only one carpel, others have several clustered together.

Ovaries
Female reproductive structures. Each is the main part of the **carpel** and contains one or more tiny bodies called **ovules***, each of which contains a female sex cell. An ovule is fixed by a stalk (**funicle**) to an area of the ovary's inside wall called the **placenta**. The stalk is attached to the ovule at a point called a **chalaza**.

Stigma
The uppermost part of a **carpel**, with a sticky surface to which grains of **pollen*** become attached during **pollination***.

Style
The part of a **carpel** which joins the **stigma** to the **ovary**. Many flowers have an obvious style, e.g. daffodils, but in others it is very short (e.g. buttercups) or almost non-existent (e.g. poppies).

Gynaecium
The whole female reproductive structure, made up of one or more **carpels**.

Buttercup carpels

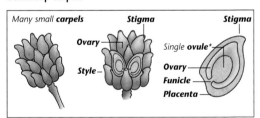

Many small **carpels** · **Stigma** · **Stigma** · **Ovary** · Single **ovule*** · **Style** · **Ovary** · **Funicle** · **Placenta**

Poppy carpel

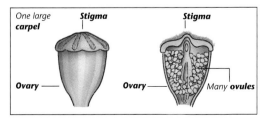

One large **carpel** · **Stigma** · **Stigma** · **Ovary** · **Ovary** · Many **ovules**

The male organs

Stamens
The male reproductive organs. Each has a thin stalk, or **filament**, with an **anther** at the tip. Each anther is made up of **pollen sacs**, which contain grains of **pollen***.

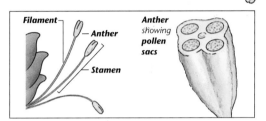

Stamens

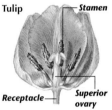

Filament · Anther · Anther showing pollen sacs · Stamen

Androecium
A collective term for all the male parts of a flower, i.e. all the **stamens**.

How the parts are arranged

Hypogynous flower
The **carpel** (or carpels) sit on top of the **receptacle**; all the other parts grow out from around its base. The position of the carpel is described as **superior**.

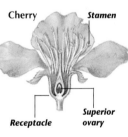

Tulip · Stamen · Receptacle · Superior ovary

Perigynous flower
The **carpel** (or carpels) rest in a cup-shaped **receptacle**; all the other parts grow out from around its rim. The position of the carpel is described as **superior**.

Cherry · Stamen · Receptacle · Superior ovary

Epigynous flower
The flower parts grow from the top of a **receptacle** which completely encloses the **ovary** (or ovaries), but not the **stigma** and **style**. The position of the ovary is described as **inferior**.

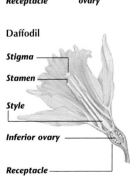

Daffodil · Stigma · Stamen · Style · Inferior ovary · Receptacle

* **Ovules, Pollen, Pollination**, 30.

REPRODUCTION IN A FLOWERING PLANT

Reproduction is the creation of new life. All flowering plants reproduce by **sexual reproduction***, when a male **gamete*** (sex cell) joins with a female gamete. In flowering plants, the male gametes (strictly speaking only male **nuclei***) are held in **pollen**, and the female gametes in **ovules**.

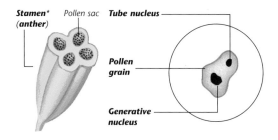

*Insects play a major part in the **cross pollination** of some plant species (see page 31).*

Pollen

Tiny grains formed by the **stamens*** (male parts) of flowers (see picture, right). Each grain is a special cell which has two **nuclei***. When a pollen grain lands on an **ovary*** (female body), one nucleus (the **generative nucleus**) divides into two, forming two **male nuclei** (reproductive bodies – see introduction).

Stamen* (anther) — *Pollen sac* — *Tube nucleus*

Pollen grain

Generative nucleus

Ovules

The tiny structures inside a flower's female body, or **ovary***. They become seeds after **fertilization**. Each consists of an oval cell (the **embryo sac**), surrounded by layers of tissue called **integuments**, except at one point where there is a tiny hole (**micropyle**). Before fertilization, the embryo sac **nucleus*** undergoes several divisions (looked at in more detail on page 95 – under **gamete production, female**). This results in a number of new cells (some of which become part of the seed's food store), and two naked nuclei which fuse together. One of the new cells is the female **gamete*** (sex cell), or **egg cell**.

Pollination

The process by which a grain of **pollen** transfers its **male nuclei** (see **pollen**) into the **ovary*** of a flower. The grain lands on the **stigma***, and forms a **pollen tube**, under the control of the **tube nucleus** (the one which did not divide – see **pollen**). The tube grows down through the ovary tissue and enters an **ovule** via its **micropyle**. The two male nuclei then travel along it.

Fertilization

After **pollination**, one **male nucleus** (see **pollen**) fuses with the **egg cell** in the **ovule** to form a **zygote*** (the first cell of a new plant). The other joins with the two fused female nuclei to form a cell which develops into the **endosperm***.

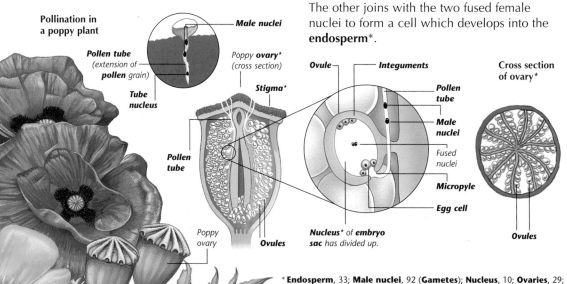

Pollination in a poppy plant

Male nuclei

Pollen tube (extension of pollen grain)

Poppy ovary* (cross section)

Tube nucleus

Stigma*

Pollen tube

Poppy ovary

Ovules

Ovule — *Integuments*

Pollen tube

Male nuclei

Fused nuclei

Micropyle

Egg cell

Nucleus* of embryo sac has divided up.

Cross section of ovary*

Ovules

* **Endosperm**, 33; **Male nuclei**, 92 (**Gametes**); **Nucleus**, 10; **Ovaries**, 29; **Sexual reproduction**, 92; **Stamens**, **Stigma**, 29; **Zygote**, 92.

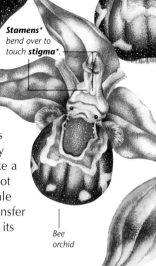

Stamens* bend over to touch **stigma***.

Cross pollination

The **pollination** of one plant by **pollen** grains from another plant of the same type. (If the grains land on a different type of plant, they do not develop further, i.e. they do not produce **pollen tubes**.) The pollen may be carried by the wind, or by insects which drink the **nectar***.

Sage flowers

Bees visit flower to drink nectar.

Pollen sticks to body of bee and is brushed off on another flower.

Self pollination

The **pollination** of a plant by its own **pollen** grains. For example, a bee orchid tries to attract male Eucera bees (for **cross pollination**) by looking and smelling like a female bee. But if it is not visited, its **stamens*** (male parts) bend over and transfer pollen to the **stigma*** of its **ovary*** (female body).

Bee orchid

Types and arrangements of flowers

Inflorescence
A group of flowers or **flowerheads** growing from one point.

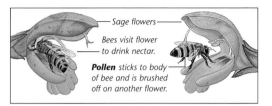

Flowering rush

Single flower

Flowerhead or composite flower
A cluster of tiny flowers, or **florets**.

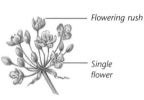

Cornflower

Florets

Umbellifer
An **inflorescence** with umbrella-shaped **flowerheads** (**umbels**).

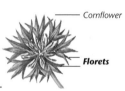

Wild carrot

Umbels

Ray florets
Florets with one long petal.

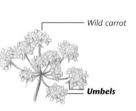

Mid-summer daisy

Ray florets

Disc florets
Florets whose petals are all the same size.

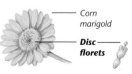

Corn marigold

Disc florets

Bell flower
Also called a **tubular** or **campanulate flower**. Its petals are joined to make a bell shape.

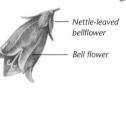

Nettle-leaved bellflower

Bell flower

Spurred flower
A flower with one or more petals extended backwards to form **spurs**.

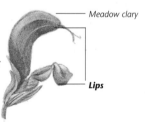

Columbine

Spurs

Lipped flower
A flower with two "lips" – an upper and lower one. The upper one often has a hood.

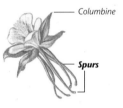

Meadow clary

Lips

Pea flower
A flower with an upper petal (the **standard**), two side petals (**wing petals**) and two lower petals forming the **keel** (which encloses the reproductive parts).

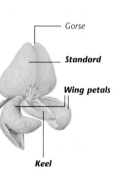

Gorse

Standard

Wing petals

Keel

*Nectar, 28 (**Nectaries**);
Ovaries, Stamens, Stigma, 29.

31

SEEDS AND GERMINATION

After **fertilization*** in a flowering plant, an **ovule*** develops into a **seed**. This contains an **embryo**, i.e. a new developing plant, and a store of food. The **ovary*** ripens into a fruit, carrying the seed or seeds. You can find out more about different fruits on page 28.

Rowan berries are **indehiscent** *and are eaten by birds.*

Dispersal or dissemination
The shedding of ripe seeds from the fruit of a parent plant. This happens in one of two main ways, depending on whether a fruit is **dehiscent** or **indehiscent**.

Dehiscent
A word describing a fruit from which the seeds are expelled before the fruit itself disintegrates. For example, a poppy capsule has holes in it, and the seeds are shaken out by the wind. Other fruits, e.g. pea pods, open spontaneously and "shoot" the seeds out. In many cases, the seeds may then be carried by wind, water or other means.

Pea pods burst open.

Poppy capsule

Seeds are shaken out.

Seeds

Indehiscent
A word describing a fruit which becomes detached from the plant and disintegrates to free the seeds. For example, the "keys" of sycamores or the "parachutes" of dandelions are carried by the air, and hooked burrs catch on animal fur. The fruit then rots away in the ground to expose the seeds. Edible fruits may be eaten by animals, which then expel the seeds in their droppings.

Strawberries are eaten by animals.

Dandelion "parachutes" are carried by the wind.

Burdock hooks catch on animals' fur.

Germination

When conditions are right, a seed will **germinate**. The **plumule** and **radicle** emerge from the seed coat, and begin to grow into the new plant, or **seedling**.

Plumule

Seed starting to **germinate**

Testa

Radicle

Hypogeal
A type of **germination**, e.g. in pea plants, in which the **cotyledons** remain below the ground within the **testa**, and the **plumule** is the only part to come above the ground.

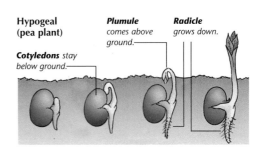

Hypogeal (pea plant)

Plumule comes above ground.

Radicle grows down.

Cotyledons stay below ground.

 ***Fertilization**, 30; **Ovaries**, 29; **Ovules**, 30.

Parts of a seed

Hilum
A mark on a seed, showing where the **ovule*** was attached to the **ovary***.

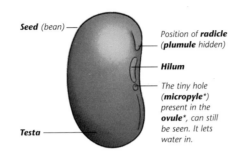

Seed (bean)

Position of **radicle** (**plumule** hidden)

Hilum

The tiny hole (**micropyle***) present in the **ovule***, can still be seen. It lets water in.

Testa

Testa
The seed coat. It develops from the **integuments***.

Plumule
The first bud, or **primary bud**, formed inside a seed. It will develop into the first shoot of the new plant.

Radicle
The first root, or **primary root**, of a new plant. It is formed inside a seed.

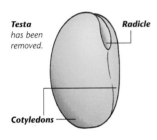

Testa has been removed.

Radicle

Cotyledons

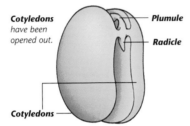

Cotyledons have been opened out.

Plumule

Radicle

Cotyledons

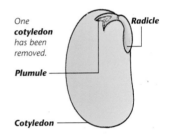

One **cotyledon** has been removed.

Radicle

Plumule

Cotyledon

Endosperm
A layer of tissue inside a seed which surrounds the developing plant and gives it nourishment. In some plants, e.g. pea, the **cotyledons** absorb and store all the endosperm before the seed is ripe; in others, e.g. grasses, it is not fully absorbed until after the seed **germinates**.

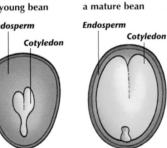

Cross section of a young bean

Endosperm

Cotyledon

Cross section of a mature bean

Endosperm

Cotyledon

Cotyledon or **seed-leaf**
A simple leaf which forms part of the developing plant. In some seeds, e.g. bean seeds, it absorbs and stores all the food from the **endosperm**. **Monocotyledons** are plants with one cotyledon, e.g. grasses; in **dicotyledons**, e.g. peas, there are two.

Epigeal
A type of **germination**, e.g. in bean plants, in which the **cotyledons** appear above the ground, below the first leaves – the true leaves.

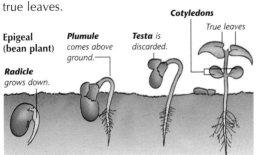

Cotyledons

True leaves

Epigeal (bean plant)

Plumule comes above ground.

Testa is discarded.

Radicle grows down.

Coleoptile
The first leaf of many **monocotyledons** (see **cotyledon**). It protects the first bud, and the first leaves emerge from it.

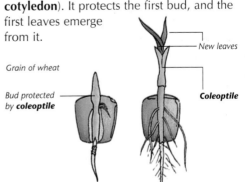

New leaves

Grain of wheat

Bud protected by **coleoptile**

Coleoptile

---

* **Integuments, Micropyle**, 30 (**Ovules**); **Ovaries**, 29.

FRUIT

A **fruit** contains the seeds of a plant. **True fruit** develop purely from the **ovary***, **false fruit** develop from the **receptacle*** as well (e.g. a strawberry). The outer wall of a fruit is called the **pericarp**. In some fruits, it is divided into an outer skin, or **epicarp**, a fleshy part, or **mesocarp**, and an inner layer, or **endocarp**. Listed below are the main types of fruit.

Grapefruit

Seed

Tomato seeds surrounded by juice

Berry

A fleshy fruit which contains many seeds, e.g. a tomato or a grapefruit. The "flesh" of citrus fruits is made up of tiny hairs, each one swollen up and full of juice.

Legume or pod

A fruit with seeds attached to its inside wall. It splits along its length to open, e.g. a pea.

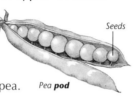

Seeds

*Pea **pod***

Grain

Also called a **caryopsis** or **kernel**. A small fruit whose wall has fused with the seed coat, e.g. wheat.

***Grains** of wheat*

Nut

A dry fruit with a hard shell, which only contains one seed, e.g. a hazelnut or a walnut.

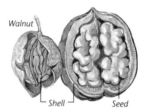

Walnut

Shell

Seed

Pome

A fruit with a thick, fleshy, outer layer and a core, with the seeds enclosed in a capsule, e.g. an apple. Pomes are examples of **false fruits** (see introduction).

Apple

Seed

Capsule

Drupe

A fleshy fruit with a hard seed in the middle, often known as a "stone", e.g. a plum.

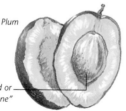

Plum

Seed or "stone"

Achene

A small, dry fruit, with only one seed, e.g. a sycamore or buttercup fruit. A "winged" achene like a sycamore fruit is a **samara** or **key fruit**.

Seed

*Sycamore **samara***

Artificial propagation

Plant cutting

Artificial propagation is the commercial process, in agriculture and market gardening, which makes use of **vegetative reproduction** (see opposite). The fact that new plants need not always grow from seeds means that many more plants can be produced commercially than would occur naturally.

Cutting

A process in which a piece of a plant stem (the cutting) is removed from its parent plant and planted in soil, where it grows into a new plant. In some cases, it is first left in water for a while to develop roots.

Taking a cutting   *Cutting in water*   *Cutting replanted*

* **Ovaries**, 29; **Receptacle**, 28.

VEGETATIVE REPRODUCTION

As well as producing seeds, some plants have developed a special type of **asexual reproduction***, called **vegetative reproduction** or **vegetative propagation**, in which one part of the plant is able to develop unaided into a new plant.

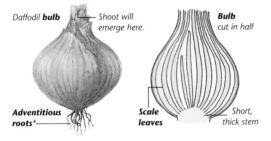

Daffodil **bulb** — Shoot will emerge here.

Bulb cut in half

Adventitious roots*

Scale leaves

Short, thick stem

Bulb

A short, thick stem surrounded by scaly leaves (**scale leaves**) which contain stored food material. It is formed underground by an old, dying plant, and represents the first, resting, stage of a new plant, which will emerge as a shoot at the start of the next growing season. E.g. a daffodil bulb (see picture above).

Corm

A short, thick stem, similar to a **bulb**, except that the food store is in the stem itself. E.g. a crocus corm.

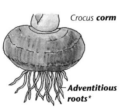

Crocus **corm**

Adventitious roots*

Rhizome

A thick stem, which has scaly leaves and grows horizontally underground. It produces roots along its length and also buds from which new shoots grow. Many grasses produce rhizomes, as well as other plants, e.g. mint and irises.

Mint **rhizome**

Rhizome cut in half

New bud

Roots

Stolon or runner

A stem which grows out horizontally near the base of some plants, e.g. the strawberry. The stolon puts down roots from points at intervals along this stem, and new plants grow at these points.

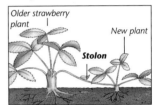

Older strawberry plant

New plant

Stolon

Tuber

A short, swollen, underground stem which contains stored food material and produces buds from which new plants will grow, e.g. a potato.

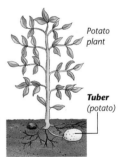

Potato plant

Tuber (potato)

Grafting

The process of removing a piece of a plant stem and re-attaching it elsewhere. The stem piece could be attached to a different part of the same plant (**autografting**), to another plant of the same species (**homografting**), or to a plant of a different species (**heterografting**). The piece removed is called the **scion**, and that to which it is attached is known as the **stock**.

Budding

A type of **grafting** where a bud and its adjacent stem are the parts grafted.

Grafting

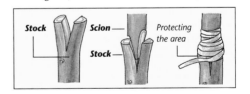

Stock

Scion

Protecting the area

Stock

Budding

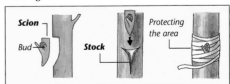

Scion

Bud

Stock

Protecting the area

* **Adventitious roots**, 17; **Asexual reproduction**, 93.

THE BODY STRUCTURE OF ANIMALS

Animals exist in a great variety of forms, from single-celled organisms to complex ones made of thousands of cells. The way they are **classified***, or divided into groups, depends to a large extent on how complex their bodies are. The two terms **higher animal** and **lower animal** are often used in this context. The higher an animal is, the more complex its internal organs are. In general, the distinguishing features of higher animals are **segmentation**, body cavities and some kind of skeleton.

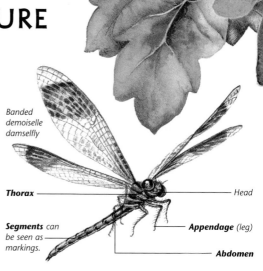

Banded demoiselle damselfly

Thorax — | — **Head**

Segments *can be seen as markings.* — | — **Appendage** (leg)

— **Abdomen**

Appendage
A subordinate body part, i.e. one which projects from the body, such as an arm, leg, fin or wing.

Segmentation
The division of a body into separate areas, or **segments**, a step up in complexity from a simple undivided body. Generally, the more complex the animal, the less obvious its segments are. The most primitive form of segmentation is **metameric segmentation**, or **metamerism**. The segments (**metameres**) are very similar, if not identical. Each contains more or less identical parts of the main internal systems, which join up through the internal walls separating the segments. Such segmentation is found in most worms, for example. More complex segmentation is less obvious. In insects, for example, the body has three main parts – the head, **thorax** (upper body region) and **abdomen** (lower body region). Each of these is in fact a group of segments, called a **tagma** (pl. **tagmata**), but the segments are not divided by internal walls. They are simply visible as external markings.

Metameric segmentation in an earthworm

— **Metamere**

Arrangement of the parts

Bilateral symmetry
An arrangement of body parts in which there is only one possible body division which will produce two mirror-image halves. It is typical of almost all freely-moving animals. The same state in flowers is called **zygomorphy** (e.g. in snapdragons).

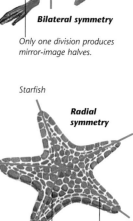

Frog

Bilateral symmetry

Only one division produces mirror-image halves.

Radial symmetry
A radiating arrangement of body parts around a central axis, e.g. in starfish. In such cases, there are two or more possible body divisions (sometimes in different planes) which will produce two mirror-image halves. The same state in flowers is called **actinomorphy** (e.g. in buttercups).

Starfish

Radial symmetry

Several different divisions produce mirror-image halves.

* **Classification**, 112.

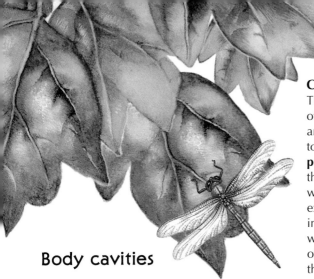

Body cavities

Almost all many-celled animals have a main fluid-filled body cavity, or **perivisceral cavity**, to cushion the body organs (very complex animals, e.g. humans, may have other smaller cavities as well). Its exact nature varies, but in most animals it is either a **coelom** or a **haemocoel**. In soft-bodied animals it is vital in movement, providing an incompressible "bag" for their muscles to work against. Such a system is called a **hydrostatic skeleton**.

Simplified cut-away of a peanut worm
(Not all body organs are shown.)

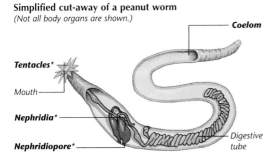

Coelom

Tentacles*

Mouth

Nephridia*

Nephridiopore*

Digestive tube

Coelom

The main body cavity (**perivisceral cavity**) of higher worms, **echinoderms***, e.g. starfish, and **vertebrates***, e.g. birds. It is fluid-filled to cushion the organs and is bounded by the **peritoneum**, a thin membrane which lines the body wall. In lower animals, e.g. many worms, the coelom assists in excretion. Their excretory organs, called **nephridia***, project into the coelom and remove fluid waste which has seeped into it. In higher animals, other more complex organs deal with these functions.

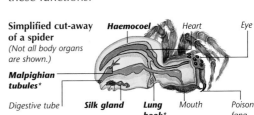

Simplified cut-away of a spider
(Not all body organs are shown.)

Haemocoel   Heart   Eye

Malpighian tubules*

Digestive tube   Silk gland   Lung book*   Mouth   Poison fang

Haemocoel

The fluid-filled main body cavity (**perivisceral cavity**) of **arthropods***, e.g. insects, and **molluscs***, e.g. snails. In molluscs, it is more of a spongy meshwork of tissue than a true cavity. Unlike a **coelom**, a haemocoel contains blood. It is an expanded part of the blood system, through which blood is circulated. In some animals, the haemocoel plays a part in excretion. In insects, for instance, water and fluid waste seep into it, and are then taken up by the **Malpighian tubules*** projecting into it.

Mantle cavity

A body cavity in shelled **molluscs***, e.g. snails. It lies between the **mantle** (a fold of skin lining the shell) and the rest of the body. Digestive and excretory waste is passed into it, for removal from the body. In water-living molluscs, it also holds the **gills***; in land-living snails, it acts as a lung.

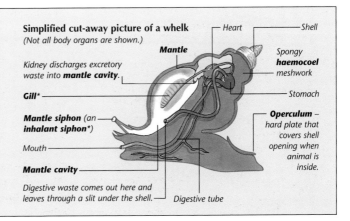

Simplified cut-away picture of a whelk
(Not all body organs are shown.)

Heart   Shell

Mantle   Spongy **haemocoel** meshwork

Kidney discharges excretory waste into **mantle cavity**.

Gill*   Stomach

Mantle siphon (an inhalant siphon*)

Operculum – hard plate that covers shell opening when animal is inside.

Mouth

Mantle cavity

Digestive waste comes out here and leaves through a slit under the shell.   Digestive tube

ANIMAL BODY COVERINGS

All animals have an enclosing outer layer, or "skin", normally with a further covering of some kind. In many cases, the skin has many layers, like human skin (see pages 82-83), and in most higher animals its covering is soft, e.g. hair, fur or feathers. Hard coverings, e.g. shells, are found in many lower animals and may form their only supporting framework, if they have no internal skeleton (**endoskeleton**). In such cases, the covering is called an **exoskeleton**. Some of the main body coverings are listed here.

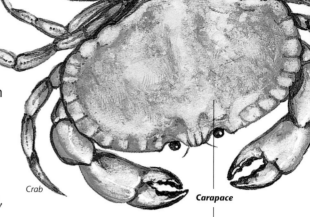

Crab

Carapace

Cuticle

Earwig

Sclerites (cuticle)

A non-living, waterproof, outer layer in many animals, secreted by the skin. In most **arthropods***, it hardens to form a supportive outer skeleton, or **exoskeleton**, e.g. the shells of crabs and the tough outer "coat" of insects. The term cuticle is in fact most often used to describe this insect "coat". It consists of a sugar-based substance (**chitin**) and a tough protein (**sclerotin**). It is often made up of **sclerites** – separate pieces joined by flexible, narrow areas. In other animals, e.g. earthworms, the cuticle remains a soft, waxy covering. (The term cuticle is sometimes used to mean the **stratum corneum*** in humans.)

Scales

There are two different types of scales. Those of bony fish, e.g. carp, are small, often bony plates lying within the skin. Those covering the limbs or whole bodies of many **reptiles*** (e.g. the legs of turtles) are thickened areas of skin.

Carapace

The shield-like shell of a crab, tortoise or turtle. In tortoises and turtles, it consists of bony plates fused together under a horny skin, but in crabs, it is a hardened **cuticle**.

Tortoise

Denticles or placoid scales

Sharp, backward-pointing plates, covering the bodies of cartilaginous fish, e.g. rays. They are similar to teeth, and stick out from the skin, unlike **scales**.

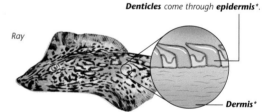

Denticles come through **epidermis***.

Ray

Dermis*

Elytra (sing. elytron)

The front pair of wings of beetles and some bugs. They are modified to form a tough cover for the back pair of wings, used for flying.

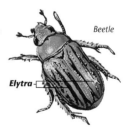

Beetle

Elytra

Scute or scutum (pl. scuta)

Any large, hard, external plate, especially those on the underside of a snake, used in movement.

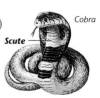

Cobra

Scute

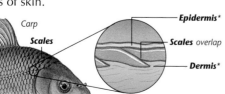

Carp

Scales

Epidermis*

Scales overlap

Dermis*

* **Arthropods**, 113; **Dermis, Epidermis**, 82; **Reptiles**, 113; **Stratum corneum**, 82.

Feathers

The insulating waterproof layer of a bird's body is made up of **feathers**, together known as its **plumage**. Each feather is a light structure made of a fibrous, horny substance called **keratin**. Each has a central **shaft** (or **rachis**) with thin filaments called **barbs**. The barbs of all **contour feathers**, i.e. all the feathers except the **down feathers**, have tiny filaments called **barbules**. Like body hairs, feathers have nerve endings attached to them, as well as muscles which can fluff them up to conserve heat (see **hair erector muscles**, page 83).

Northern parula warbler

*Rectrices (sing. **rectrix**) – tail feathers. They control changes of direction in flight.*

*Uropygium contains **uropygial gland**, which secretes an oily fluid used in preening.*

*The feathers of the back, shoulders and wings are sometimes called the **mantle**.*

*Birds' feet tend not to be covered in feathers, but are protected by small scales called **scutella** (sing. **scutellum**).*

Mandibles – upper and lower beak parts

Coverts – feathers covering bases of wing and tail feathers

*Primaries (furthest from the body) – make up end section of wing (**pinion**).*

Secondaries (nearer the body)

Remiges (sing. remix) or flight feathers
Those feathers of a bird's wings which are used in flight, consisting of the long, strong **primary feathers**, or **primaries**, and the shorter **secondary feathers**, or **secondaries**.

Down feathers or plumules
The fluffy, temporary feathers of all young birds, which have flexible **barbs**, but no true **barbules**. The adults of some types of bird keep some down feathers as an insulating layer close to the skin.

Feather follicles
Tiny pits in a bird's skin. Each one has a feather in it, just like a hair in a **hair follicle***. The cells at the base of the follicle grow up and out to form a feather, and then die away, leaving their hard, tough remains.

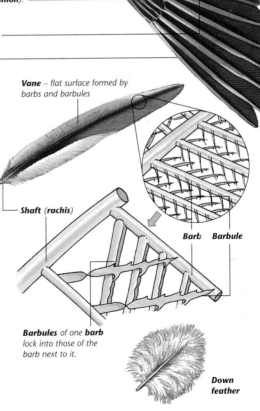

Vane – flat surface formed by barbs and barbules

Shaft (rachis)

Barb   Barbule

*Barbules of one **barb** lock into those of the barb next to it.*

Down feather

ANIMAL MOVEMENT

Most animals are capable of movement from place to place (**locomotion**) at least at some stage of their life (plants can only move individual parts – see **tropism**, page 23). The moving parts of animals vary greatly. Many animals have a system of bones and muscles similar to humans (see pages 50-55). Some of the parts used in animal movement are shown on these pages.

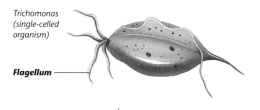

*All fish use **fins** to help them move from one place to another.*

Movement of simple animals

Pseudopodium (pl. pseudopodia)

An extension ("false foot") of the cell matter, or **cytoplasm***, of a single-celled organism. Such extensions are formed either in order for the organism to move or to enable it to engulf a food particle. The latter process is called **phagocytosis**.

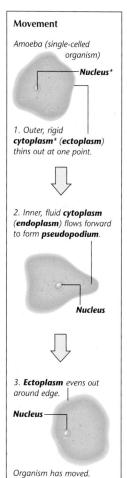

Movement

Amoeba (single-celled organism)

— **Nucleus***

1. Outer, rigid **cytoplasm*** (**ectoplasm**) thins out at one point.

2. Inner, fluid **cytoplasm** (**endoplasm**) flows forward to form **pseudopodium**.

Nucleus

3. **Ectoplasm** evens out around edge.

Nucleus —

Organism has moved.

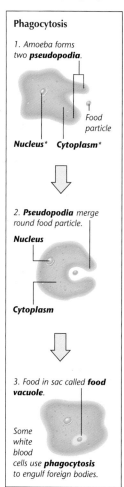

Phagocytosis

1. Amoeba forms two **pseudopodia**.

Food particle

Nucleus* **Cytoplasm***

2. **Pseudopodia** merge round food particle.

Nucleus

Cytoplasm

3. Food in sac called **food vacuole**.

Some white blood cells use **phagocytosis** to engulf foreign bodies.

Cilia (sing. cilium)

Tiny "hairs" on the outer body surfaces of many small organisms. They flick back and forth to produce movement. Cilia are also found lining the internal passages of more complex animals, e.g. human air passages (where they trap foreign particles).

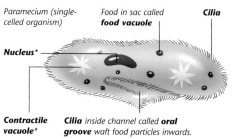

Paramecium (single-celled organism) Food in sac called **food vacuole** **Cilia**

Nucleus*

Contractile vacuole* **Cilia** inside channel called **oral groove** waft food particles inwards.

Flagella (sing. flagellum)

Any long, fine body threads, especially the one or more which project from the surface of many single-celled organisms. These lash backwards and forwards to produce movement. Organisms with flagella are **flagellate**.

Trichomonas (single-celled organism)

Flagellum —

Parapodia (sing. parapodium)

Paired projections from the sides of many aquatic worms, used to move them along. Each one ends in a bunch of bristles, or **chaetae** (sing. **chaeta**), which may also cover the body in some cases.

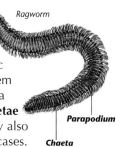

Ragworm

Parapodium

Chaeta

* **Contractile vacuoles**, 45; **Cytoplasm, Nucleus**, 10.

Swimmers

Fins

Projections from the body of a fish, which are used as stabilisers and to change direction. They are supported by **rays** – rods of bone or **cartilage*** (depending on the type of fish) radiating out inside them. Fish have two sets of fins, called **median** and **paired fins**.

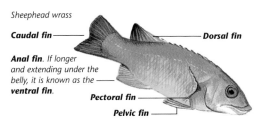

Sheephead wrass

Caudal fin ———————— Dorsal fin

Anal fin. If longer and extending under the belly, it is known as the —— **ventral fin**.

Pectoral fin

Pelvic fin

*In many fish, the **pelvic fins** are behind the **pectoral fins**.*

Median fins

The **fins** which run in a line down the centre of the back and the belly. In some fish, e.g. eels, they form one long, continuous median fin, but in most they are divided into the **dorsal**, **caudal** (tail) and **anal** (or **ventral**) fins. The dorsal fin and the anal fin control changes of direction from side to side. The caudal fin helps to propel the fish through the water.

Paired fins

The **fins** of a fish which stick out from its sides in two pairs: the **pectoral fins** and the **pelvic fins**. They control movement up or down.

Swim bladder or air bladder

A long, air-filled pouch inside most bony fish. The fish alters the amount of air inside the bladder depending on the depth at which it is swimming. This keeps the density of the fish the same as that of the water, so it will not sink if it stops swimming.

Swim bladder

Median or **medial** means "lying on the dividing line between the right and left sides".	**Caudal** means "of the tail or hind part"; **caudate** means "having a tail".
Dorsal means "of the back or top surface".	**Ventral** means "of the front or lower surface".

Flyers

Pectoralis muscles

Two large, paired chest muscles, found in many **mammals***, but especially highly developed in birds. Each wing has one **pectoralis major** and one **pectoralis minor**, attached at one end to the **keel**, a large extension of the breastbone. The muscles contract alternately to move the wings.

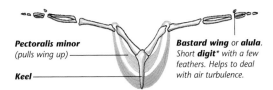

Pectoralis minor (pulls wing up) ———

Keel ———

Bastard wing or **alula**. Short **digit*** with a few feathers. Helps to deal with air turbulence.

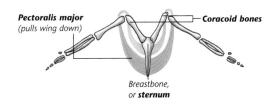

Pectoralis major (pulls wing down) ———

Coracoid bones

*Breastbone, or **sternum***

Walkers

Unguligrade
Walking on hoofs at the tips of the toes, e.g. horses.

Digitigrade
Walking on the underside of the toes, e.g. dogs and cats.

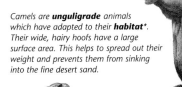

Plantigrade
Walking on the underside of the whole foot, e.g. humans.

*Camels are **unguligrade** animals which have adapted to their **habitat***. Their wide, hairy hoofs have a large surface area. This helps to spread out their weight and prevents them from sinking into the fine desert sand.*

ANIMAL FEEDING

Different animals take in their food in many different ways, and with many different body parts. Some also have special internal mechanisms for dealing with the food (others have human-like **digestive systems** – see pages 66-67). Listed here are some of the main animal body parts involved in feeding and digestion.

Sea anemone

Cnidoblasts or thread cells

Special cells found in large numbers on the **tentacles*** of **cnidarians***, e.g. sea anemones, used for seizing food. Each one contains a **nematocyst** – a long thread coiled inside a tiny sac. When a tentacle touches something, the threads shoot out to stick to it or sting it.

Cut-away
*tentacle***

Cnidoblast

Nematocyst shoots out

Filter-feeding

The "sieving" of food from water, shown by many aquatic animals. Barnacles, for instance, sieve out microscopic organisms, or **plankton***, with bristly limbs called **cirri** (sing. **cirrus**).

Cirri

*Barnacles put out their **cirri** when they are covered with water.*

Diastema (pl. diastemata)

A gap between the front and back teeth of many plant-eaters. It is especially important in rodents, e.g. mice. They can draw their cheeks in through the gaps, so they do not swallow substances they may be gnawing.

Some whales use frayed plates of horny **baleen**, or **whalebone**, hanging down from the top jaw. They sieve out small, shrimp-like animals called **krill**.

Krill

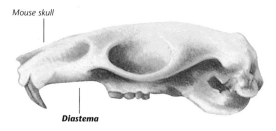

Mouse skull

Diastema

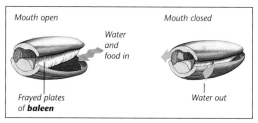

Mouth open

Water and food in

Mouth closed

*Frayed plates of **baleen***

Water out

Carnassial teeth

The specially adapted last upper **premolar*** and first lower **molar*** of carnivorous mammals, used for shearing flesh.

Radula

The horny "tongue" of many **molluscs***, e.g. snails. It is covered by tiny teeth, which rasp off food.

*A grey whale filtering sea water through its **baleen***

Arthropod mouthparts

The mouths of **arthropods***, e.g. insects, are made up of a number of different parts. Depending on the animal's feeding method, these may look very different. The basic mouthparts, found in all insects, are the **mandibles, maxillae** (sing. **maxilla**), **labrum** and **labium**. The first two are also found in many other arthropods, e.g. crabs and centipedes (some of these other arthropods have two pairs of maxillae).

*The **maxillae** of butterflies, moths and similar insects fit together to make a long sucking tube, or **proboscis**.*

*The **labium** of houseflies is an extended pad-like sucking organ.*

*Grooves called **pseudotracheae** (sing. **pseudotrachea**)*

Typical arrangement of mouthparts (locust)

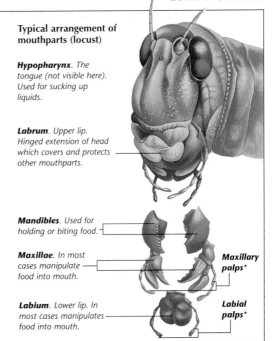

Hypopharynx. *The tongue (not visible here). Used for sucking up liquids.*

Labrum. *Upper lip. Hinged extension of head which covers and protects other mouthparts.*

Mandibles. *Used for holding or biting food.*

Maxillae. *In most cases manipulate food into mouth.*

Maxillary palps*

Labium. *Lower lip. In most cases manipulates food into mouth.*

Labial palps*

Digestive structures

Crop

A thin-walled pouch, part of the gullet (**oesophagus***) in birds; also a similar structure in some worms, e.g. earthworms, and some insects, e.g. grasshoppers. Food is stored in the crop before it goes into the **gizzard**.

Gizzard

A thick, muscular-walled pouch at the base of the gullet (**oesophagus***) in those animals which have **crops**. These animals have no teeth, instead food is ground up in the gizzard. Birds swallow pieces of gravel to act as grindstones; in other animals, the muscular walls of the gizzard, or hard, tooth-like structures attached to these walls, help to grind up their food.

Pigeon

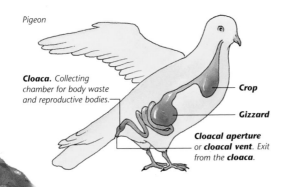

Cloaca. *Collecting chamber for body waste and reproductive bodies.*

Crop

Gizzard

Cloacal aperture *or **cloacal vent**. Exit from the **cloaca**.*

Rumen

The large first chamber of the complex "stomach" of some plant-eating **mammals***, e.g. cows, into which food passes unchewed. It contains bacteria which can break down **cellulose***. Other animals pass this substance as waste, but these animals cannot afford to do this, as it makes up the bulk of their food (grass). Partially-digested food from the rumen is digested further in the second chamber, or **reticulum**, and then regurgitated to be chewed, when it is known as the **cud**. When swallowed again, it passes directly to the third and fourth chambers – the **omasum** and the **abomasum** (the true stomach) – for further processing.

*Animals that chew the cud (**ruminate**) are called **ruminants**.*

To the intestines

Abomasum

Reticulum

Rumen

Omasum

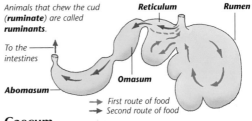

→ *First route of food*
→ *Second route of food*

Caecum

Any blind-ended sac inside the body, especially one forming part of a digestive system. In many animals, e.g. rabbits, it is the site of an important stage of digestion (involving bacterial breakdown of **cellulose*** – see **rumen**). In others, e.g. humans (see **large intestine***), it is redundant.

* **Arthropods**, 113; **Cellulose**, 10 (**Cell wall**); **Large intestine**, 67; **Mammals**, 113; **Oesophagus**, 66; **Palps**, 46.

ANIMAL RESPIRATION

The complex process of **respiration** consists of a number of stages (see introduction, page 70). Basically, oxygen is taken in and used by body cells in the breakdown of food, and carbon dioxide is expelled from the cells and the body. On these pages are some of the main animal respiratory organs.

These narrow slits are where water leaves, having passed over the shark's **gills**.

Gills

Gills or **branchiae** (sing. **branchia**), are the breathing organs of most aquatic animals, containing many blood vessels. Oxygen is absorbed into the blood from the water passing over the gills. Carbon dioxide passes out the other way. There are two types of gills – **internal** and **external**.

Breathing with gills

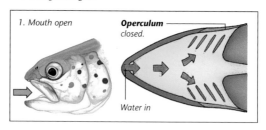

1. Mouth open

Operculum closed.

Water in

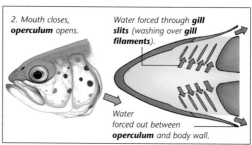

2. Mouth closes, **operculum** opens.

Water forced through **gill slits** (washing over **gill filaments**).

Water forced out between **operculum** and body wall.

Internal gills

Gills inside the body, found in various forms in all fish, most **molluscs***, e.g. limpets, and most **crustaceans** (a group of **arthropods*** which includes crabs). Most fish have four pairs of gills, with openings between them called **gill slits**. In more advanced fish, e.g. cod, they are covered by a flap called the **operculum**. In more primitive fish, e.g. sharks, they end in narrow openings in the skin on the side of the head. Each gill consists of a curved rod, the **gill bar** or **gill arch**, with many fine **gill filaments**, and even finer **gill lamellae** (sing. **lamella**) radiating from it. These all contain blood vessels.

External gills

Gills on the outside of the body, found in the young stages of most fish and **amphibians***, some older amphibians and the young aquatic stages of many insects (e.g. caddisfly **larvae*** and mayfly **nymphs***). Their exact form depends on the type of animal, but in many cases they are "frilly" outgrowths from the head, e.g. in young tadpoles.

Tadpole

External gills are soft and "frilly".

Siphon

A tube carrying water to (**inhalant siphon**) or from (**exhalant siphon**) the **gills** of many lower aquatic animals, e.g. whelks (see picture, page 37). The exhalant siphon of **cephalopods** (**molluscs*** with **tentacles***), e.g. octopuses, is called the **hyponome***.

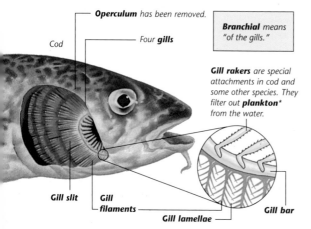

Operculum has been removed.

Cod

Four **gills**

Branchial means "of the gills."

Gill rakers are special attachments in cod and some other species. They filter out **plankton*** from the water.

Gill slit

Gill filaments

Gill lamellae

Gill bar

Other respiratory organs

Spiracle

Any body opening through which oxygen and carbon dioxide are exchanged (e.g. a whale's blowhole). The term is used especially for any of the tiny holes (also called **stigmata**, sing. **stigma**) found in many **arthropods***, e.g. insects.

Tracheae (sing. **trachea**[†])

Thin tubes leading in from the **spiracles** of all insects (and the most advanced spiders). They form an inner network, often branching into narrower tubes called **tracheoles**. Oxygen from the air passes through the tube walls to the body cells. Carbon dioxide leaves via the same route.

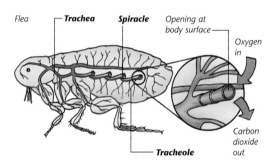

Flea — Trachea — Spiracle — Opening at body surface — Oxygen in — Carbon dioxide out — Tracheole

Lung books or book lungs

Paired breathing organs found in scorpions (which have four pairs) and some (less advanced) spiders (which have one or two). Each one has many blood-filled tissue plates, arranged like book pages. Oxygen comes in through slits (**spiracles**), one by each lung book, and is absorbed into the blood. Carbon dioxide passes out the same way.

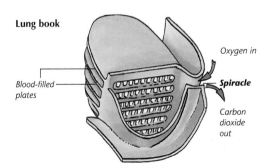

Lung book — Blood-filled plates — Oxygen in — Spiracle — Carbon dioxide out

Animal excretion

Excretion

The expulsion of waste fluid. It is vital to life as it gets rid of harmful substances. It is also vital to the maintenance of a balanced level of body fluids (see **homeostasis**, page 107).

Contractile vacuoles

Tiny sacs used for water-regulation in single-celled freshwater organisms. Excess water enters a vacuole via several canals arranged around it. At intervals, the vacuole expels its contents.

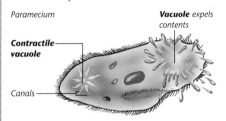

Paramecium — Vacuole expels contents — Contractile vacuole — Canals

Nephridia (sing. **nephridium**)

Waste-collecting tubes in many worms and the **larvae*** of many **molluscs***, e.g. mussels. In higher worms, they collect from the **coelom*** (see picture, page 37). Lower worms and mollusc larvae have more primitive **protonephridia**. The waste fluid enters these via hollow **flame cells** (**solenocytes**), which contain hair-like **cilia***. In both a nephridium and a protonephridium, the waste leaves through a tiny hole, or **nephridiopore**.

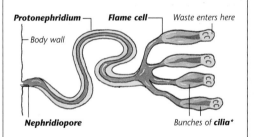

Protonephridium — Flame cell — Waste enters here — Body wall — Nephridiopore — Bunches of cilia*

Malpighian tubules

Long tubes found in many **arthropods***, e.g. insects. They carry dissolved waste from the main body cavity (**haemocoel***) into the rear of the gut. See picture, page 37.

---

* **Arthropods**, 113; **Cilia**, 40; **Coelom**, **Haemocoel**, 37; **Larva**, 49; **Molluscs**, 113.
[†] This is also the word for the human windpipe. See page 70.

ANIMAL SENSES AND COMMUNICATION

All animals show some **sensitivity** (**irritability**), i.e. response to external stimuli such as light and sound vibrations. Humans have a high overall level of sensory development, but individual senses in other animals may be even better developed, e.g. the acute vision of hawks. Listed here are some of the main animal sense organs (and their parts). Their responding parts send "messages" (nervous impulses) to the brain (or more primitive nerve centre), which initiates the response.

Touch, smell and taste

Antennae (sing. antenna)
Whip-like, jointed sense organs on the heads of insects, centipedes and millipedes, and all **crustaceans** (a group of **arthropods*** which includes crabs and prawns). Insects, centipedes and millipedes have one pair, crustaceans have two. They respond to touch, temperature changes and chemicals (giving "smell" or "taste"). Some crustaceans also use them for swimming or to attach themselves to objects or other animals.

*The Lion's Mane jellyfish uses its **tentacles** to sting and catch fish. Some large Lion's Mane jellyfish have tentacles that are up to 30m long.*

Tentacles
Long, flexible body parts, found in many **molluscs***, e.g. octopuses, and **cnidarians***, e.g. jellyfish. In most cases they are used for grasping food or feeling, though the longer of the two pairs found in land snails and slugs have eyes on the end.

Octopus

Tentacles

Hyponome. Octopus shoots water out of it to move by "jet propulsion".

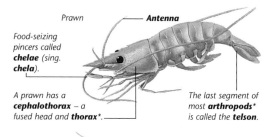

Prawn — **Antenna**

Food-seizing pincers called **chelae** *(sing.* **chela***).*

A prawn has a **cephalothorax** *– a fused head and* **thorax***.*

The last segment of most **arthropods*** *is called the* **telson***.*

Vibrissae (sing. vibrissa) or whiskers
Stiff hairs standing out from the faces of many **mammals***, e.g. cats, round the nose. They are sensitive to touch.

Palps
Projections of the mouthparts of **arthropods***, e.g. insects (see cricket picture, left). They respond to chemicals (giving "smell" or "taste"). The term is also given to various touch-sensitive organs.

— **Antenna**

Palps

Crickets have a single pair of long antennae.

Setae

Tympanal organ *(see page 47)*

Setae (sing. seta)
Bristles produced by the skin of many **invertebrates***, e.g. insects. Nerves at their bases respond to movements of air or vibrations.

* **Arthropods, Cnidarians, Invertebrates, Mammals, Molluscs**, 113; **Thorax**, 36 (**Segmentation**).

Hearing and balance

Lateral lines

Two water-filled tubes lying along each side of the body, just under the skin. They are found in all fish and those **amphibians*** which spend most of their time in water, e.g. some toads. They enable the animals to detect water currents and pressure changes, and they use this information to find their way about.

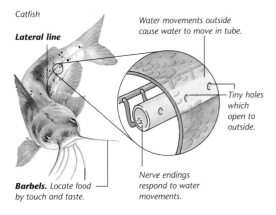

Catfish

Lateral line

Water movements outside cause water to move in tube.

Tiny holes which open to outside.

Barbels. *Locate food by touch and taste.*

Nerve endings respond to water movements.

Tympanal organs or tympani (sing. tympanum)

Sound detectors found on the lower body or legs in some insects, e.g. crickets, and on the head in some land **vertebrates***, e.g. frogs. Each is an air sac covered by a thin layer of tissue. Sensitive fibres in the organs respond to high-frequency sound vibrations.

Tympanal organ *located on side of head*

Frog

Statocysts

Tiny organs of balance, found in many aquatic **invertebrates***, e.g. jellyfish. Each is a sac with tiny particles called **statoliths** inside, e.g. sand grains. When the animal moves, the grains move, stimulating sensitive cells which set off responses.

Haltères

Modified second pair of wings on some insects, which keep balance in flight.

Fly

Haltères

Sight

Compound eyes

The special eyes of many insects and some other **arthropods***, e.g. crabs. Each consists of hundreds of separate visual units called **ommatidia** (sing. **ommatidium**). Each of these has an outer lens system which "bends", or refracts, light onto a **rhabdom**, a transparent rod surrounded by light-responsive cells.

Compound eye view of a flower (mosaic image)

Compound eye

Facet *(surface of a lens system)*

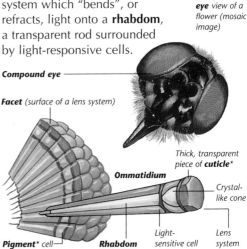

*Thick, transparent piece of **cuticle****

Ommatidium

Crystal-like cone

Light-sensitive cell

Lens system

Pigment* **cell**

Rhabdom

After receiving information from all the ommatidia (each has a slightly different angle of vision so may record different light intensity or colour), the brain assembles a complete **mosaic image**. This is enough for the animal's needs, but not as well defined as the image made by the human eye.

Communication

Pheromone

Any chemical made by an animal that causes responses in other members of the species, e.g. sexual attractants produced by many insects.

Stridulation

The rubbing together of body parts to make a shrill noise (often used to attract a mate). Crickets use their wing edges.

Syrinx (pl. syringes)

The vocal organ of birds, similar to the **larynx***, but found at the windpipe's base.

---

* **Amphibians, Arthropods**, 113; **Cuticle**, 38; **Invertebrates**, 113; **Larynx**, 70; **Pigments**, 27; **Vertebrates**, 113.

ANIMAL REPRODUCTION AND DEVELOPMENT

Reproduction is the creation of new life. Most animals reproduce by **sexual reproduction***, the joining of a female sex cell, called an **ovum**, with a male sex cell, or **sperm**. Below are the main terms associated with the reproductive processes of animals.

*Chick hatching from a **cleidoic egg***

Viviparous

A term describing animals such as humans, in which both the joining of the male and female sex cells (**fertilization**) and the development of the **embryo*** occur inside the female's body (the fertilization is **internal fertilization**), and the baby is born live.

New-born piglets suckling. Pigs are **viviparous**.

Oviparous

A term describing animals in which the development of the **embryo*** occurs in an **egg** which has been laid by the mother. In some cases, e.g. in birds, the male and female sex cells join inside the female's body (**internal fertilization**) and the egg already contains the embryo when laid. In other cases, e.g. in many fish, the many eggs each just contain an **ovum** (female sex cell) when laid, and the male then deposits **sperm** (male sex cells) over them (**external fertilization**).

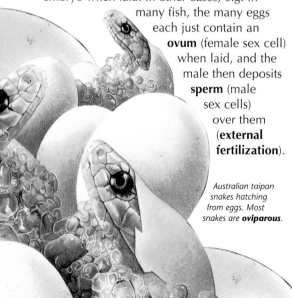

Australian taipan snakes hatching from eggs. Most snakes are **oviparous**.

Eggs

There are two main types of egg. **Cleidoic eggs** are produced by most egg-laying animals which live on land, e.g. birds and most **reptiles***, and also by a few aquatic animals, e.g. sharks. Such an egg largely isolates the **embryo*** from its surroundings, allowing only gases to pass through its tough shell (waste matter is stored). It contains enough food (**yolk**) for the complete development of the embryo, and the animal emerges as a tiny version of the adult. The other type of egg, produced by most aquatic animals, e.g. most fish, has a soft outer membrane, through which water and waste matter (as well as gases) can pass. The emerging young are not fully developed.

Cleidoic egg

Yolk (rich in phosphorus and fat). Gradually absorbed by **embryo**, together with surrounding **yolk sac** (human embryos have the remains of a yolk sac attached to them).

Albumen. The "white" of the egg – providing protein and water.

Amnion. Thin layer of tissue, making **amniotic sac** which contains cushioning **amniotic fluid**.

Chalazae (sing. **chalaza**). Twisted bands of **albumen**, holding **yolk** in place and acting as shock absorbers.

Shell

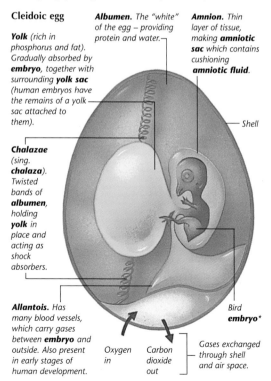

Allantois. Has many blood vessels, which carry gases between **embryo** and outside. Also present in early stages of human development.

Oxygen in

Carbon dioxide out

Gases exchanged through shell and air space.

Bird **embryo***

* **Embryo**, 92; **Reptiles**, 113; **Sexual reproduction**, 92.

Oviduct

Any tube in females through which either **eggs** or **ova** (female sex cells) are discharged to the outside. In some animals, e.g. birds, the eggs are **fertilized** on the way out (see **oviparous**).

Ovipositor

An organ extending from the back end of many female insects, through which **eggs** are laid. In many cases, it is long and sharp, and is used to pierce plant or animal tissues before laying.

Spermatheca

A sac for storing **sperm** (male sex cells) in the female of many **invertebrates***, e.g. insects, and some lower **vertebrates***, e.g. newts. The female receives the sperm and stores them until her **ova** (sex cells) are ready to join with them (**fertilization**).

*Some **hermaphrodite** animals (animals with both male and female organs), e.g. earthworms, have **spermathecae**. They "swap" sperm when they mate.*

Metamorphosis

The growth and development of some animals involves intermediate forms which are very different from the adult form. Metamorphosis is a series of such changes, producing a complete or partial transformation from the young form to the adult.

All insects, most marine **invertebrates***, e.g. lobsters, and most **amphibians***, e.g. frogs, undergo some degree of metamorphosis (intermediate larval forms are common, e.g. legless **tadpoles** in frogs and toads). Below are examples of insect metamorphosis (two different kinds – **complete** and **incomplete** metamorphosis).

*Complete metamorphosis (two different forms between **egg** and adult). The many insects which undergo it, e.g. butterflies, are called **endopteryogotes**.*

Male and female spotted fritillary butterflies of southern Europe pair up and mate. Female butterfly lays her eggs on small plant.

*Larva (pl. **larvae**) emerges from egg. Has other names, e.g. **grub** (beetles), **maggot** (houseflies), **caterpillar** (moths and butterflies). Sheds skin several times to allow for growth (process called **ecdysis**, common to all **arthropods***).*

*Final ecdysis (see **larva**) results in **pupa** (pl. **pupae**). Called **chrysalis** in butterflies. Outer skin is a hard protective case. The cases surrounding moth pupae have extra protection in the form of a **cocoon** of spun silk.*

*Hard case splits and mature adult (**imago**) emerges. The imago searches for a mate, and the reproductive cycle is then repeated.*

*Incomplete metamorphosis (gradual development in stages). The insects which undergo it, e.g. locusts, are called **exopterygotes**.*

Nymph

*Locust **nymphs** are called **hoppers**.*

Nymph. Emerges from egg. "Mini" version of adult insect, but resemblance only superficial, e.g. wings in very early stages of development or non-existent, many inner organs missing.

*Nymph undergoes several **ecdyses** (see **larva**), with some adult parts emerging each time.*

Old skin

*After last ecdysis, mature adult (**imago**) emerges.*

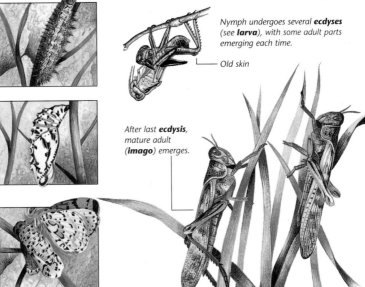

THE SKELETON

The human **skeleton** is a frame of over 200 bones which supports and protects the body organs (the **viscera**) and provides a solid base for the muscles to work against.

Cranium or skull

A case protecting the brain and facial organs. It is made of **cranial** and **facial bones**, fused at lines called **sutures**. The upper jaw, for instance, consists of two fused bones called **maxillae** (sing. **maxilla**).

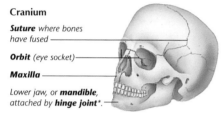

Cranium

Suture where bones have fused

Orbit (eye socket)

Maxilla

Lower jaw, or **mandible**, attached by **hinge joint**.

Rib cage

A cage of bones forming the walls of the **thorax** or chest area. It is made up of 12 pairs of **ribs**, the **thoracic vertebrae** and the **sternum**. The ribs are joined to the sternum by bands of **cartilage*** called **costal cartilage**, but only the first seven pairs join it directly. The last five pairs are **false ribs**. The top three of these join the sternum indirectly – their costal cartilage joins that of the seventh pair. The bottom two pairs are **floating ribs**, only attached to the thoracic vertebrae at the back.

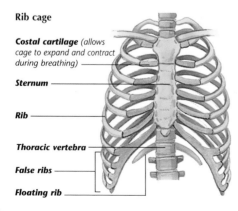

Rib cage

Costal cartilage (allows cage to expand and contract during breathing)

Sternum

Rib

Thoracic vertebra

False ribs

Floating rib

The bones of the skeleton

Seven **cervical vertebrae** support the neck. The top two are the **atlas** and **axis**.

Scapula or **shoulderblade**

Sternum or **breastbone**

Ribs

12 **thoracic vertebrae** support the **ribs**.

The five **lumbar vertebrae** are in the lower back (**lumbar**) region.

The five **sacral vertebrae** at the base of the column are fused together to form the **sacrum**.

Coccyx. An area of four fused **coccygeal vertebrae** below the **sacrum**.

Pelvis, **pelvic girdle** or **hip girdle**. Each side is made up of three bones – the **ilium**, **pubis** and **ischium**.

Tarsals (ankle bones), collectively called a **tarsus**.

Phalanges (sing. **phalanx**). The bones of the **digits** – fingers and toes.

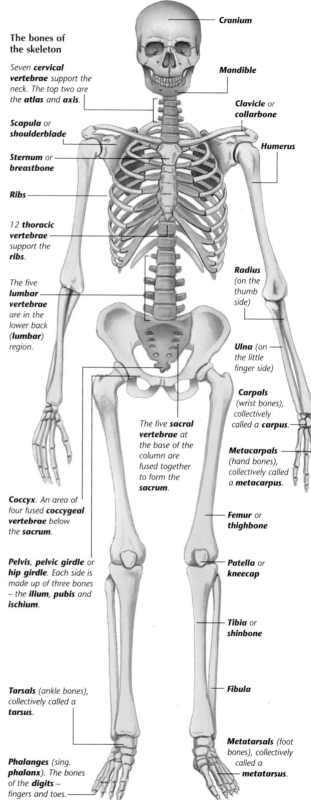

Cranium

Mandible

Clavicle or **collarbone**

Humerus

Radius (on the thumb side)

Ulna (on the little finger side)

Carpals (wrist bones), collectively called a **carpus**.

Metacarpals (hand bones), collectively called a **metacarpus**.

Femur or **thighbone**

Patella or **kneecap**

Tibia or **shinbone**

Fibula

Metatarsals (foot bones), collectively called a **metatarsus**.

* **Cartilage**, 53; **Hinge joints**, 52.

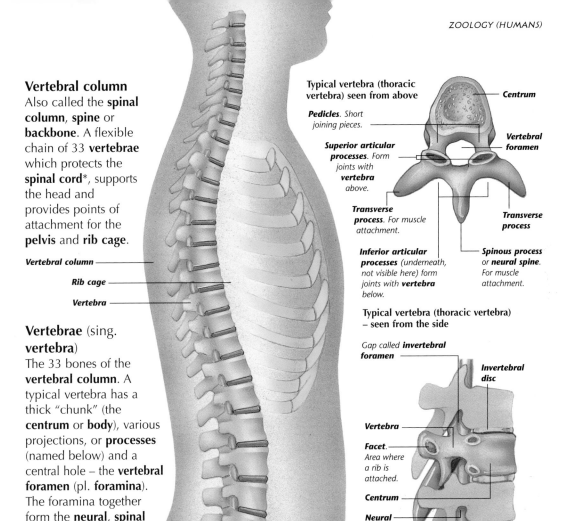

Vertebral column

Also called the **spinal column**, **spine** or **backbone**. A flexible chain of 33 **vertebrae** which protects the **spinal cord***, supports the head and provides points of attachment for the **pelvis** and **rib cage**.

Vertebral column
Rib cage
Vertebra

Vertebrae (sing. vertebra)

The 33 bones of the **vertebral column**. A typical vertebra has a thick "chunk" (the **centrum** or **body**), various projections, or **processes** (named below) and a central hole – the **vertebral foramen** (pl. **foramina**). The foramina together form the **neural**, **spinal** or **vertebral canal**, through which the **spinal cord*** runs.

Typical vertebra (thoracic vertebra) seen from above

Centrum

Pedicles. Short joining pieces.

Vertebral foramen

Superior articular processes. Form joints with **vertebra** above.

Transverse process. For muscle attachment.

Transverse process

Inferior articular processes (underneath, not visible here) form joints with **vertebra** below.

Spinous process or **neural spine**. For muscle attachment.

Typical vertebra (thoracic vertebra) – seen from the side

Gap called **invertebral foramen**

Invertebral disc

Vertebra

Facet. Area where a rib is attached.

Centrum

Neural canal

Spinal cord*

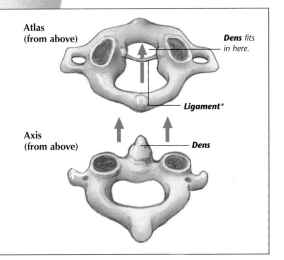

Vertebral structure

The different vertebrae are named around the skeleton on the opposite page. The top 24 are movable and linked by **invertebral discs** of **cartilage***. The bottom nine are fused together. They all have the typical structure described above right, except for the top two, the **atlas** and **axis**. The atlas (top vertebra) has a special joint with the **skull** which allows the head to nod. The axis has a "peg" (the **dens** or **odontoid process**) which fits into the atlas. This forms a **pivot joint**, a type of joint which allows the head to rotate.

Atlas (from above)

Dens fits in here.

Ligament*

Axis (from above)

Dens

JOINTS AND BONE

The bones of the skeleton meet at many **joints**, or **articulations**. Some are **fixed joints**, allowing no movement, e.g. the **sutures*** of the skull. Most, however, are movable, and they give the body great flexibility. The most common are listed on this page.

Hinge joints

Joints (e.g. the knee joint) which work like any hinge. That is, the movable part (bone) can only move in one plane, i.e. in either of two opposing directions.

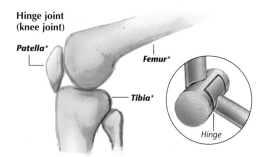

Hinge joint (knee joint)

*Patella**

*Femur**

*Tibia**

Hinge

Gliding joints

Also called **sliding** or **plane joints**. Joints in which one or more flat surfaces glide over each other, e.g. those between the **carpals***. They are more flexible than **hinge joints**.

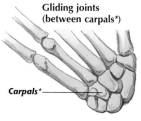

Gliding joints (between carpals*)

*Carpals**

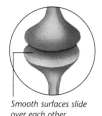

Smooth surfaces slide over each other.

Ball-and-socket joints

The most flexible joints (e.g. the hip joint). The movable bone has a rounded end which fits into a socket in the fixed bone. The movable bone can swivel, or move in many directions.

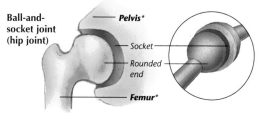

Ball-and-socket joint (hip joint)

*Pelvis**

Socket

Rounded end

*Femur**

Connective tissue

There are many different types of **connective tissue** in the body. They all protect and connect cells or organs and have a basis of non-living material (the **matrix**) in which living cells are scattered. The difference between them lies in the nature of this material. The various types of tissue found at a joint, including **bone** itself, are all types of connective tissue. They all contain protein fibres and are either tough (containing **collagen** fibres) or elastic (containing **elastin** fibres).

The ease with which different types of tissue grow and repair depends largely on the amount of blood they contain. **Periosteum** is **vascular** (has a blood supply) and repairs itself quickly. **Cartilage** is **avascular** (has no blood supply) so takes longer to repair.

Periosteum

A thin layer of elastic connective tissue. It surrounds all bones, except at the joints (where **cartilage** takes over), and contains **osteoblasts** – cells which make new bone cells, needed for growth and repair.

Ligaments

Bands of connective tissue which hold together the bones of joints (and also hold many organs in place). Most are tough, though some are elastic, e.g. between **vertebrae***.

Synovial sac or synovial capsule

A cushioning "bag" of lubricating fluid (**synovial fluid**), with an outer skin (**synovial membrane**) of elastic connective tissue. Most movable joints, e.g. the knee, have such a sac lying between the bones. They are known as **synovial joints**.

* **Carpals, Femur, Patella, Pelvis**, 50; **Sutures**, 50 (**Cranium**); **Tibia**, 50; **Vertebrae**, 51.

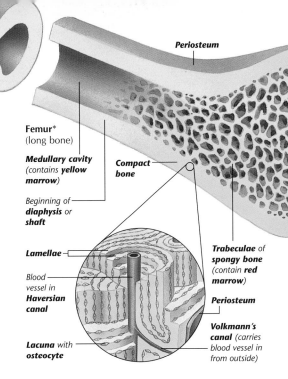

Periosteum

Femur*
(long bone)

Medullary cavity
(contains **yellow marrow**)

Compact bone

Beginning of **diaphysis** or **shaft**

Lamellae

Blood vessel in **Haversian canal**

Lacuna with **osteocyte**

Trabeculae of spongy bone (contain **red marrow**)

Periosteum

Volkmann's canal (carries blood vessel in from outside)

Spongy bone

A type of **bone** that is found in short and/or flat bones, e.g. the **sternum***, and fills the ends of long bones, e.g. the **femur***. It consists of a criss-cross network of flat plates called **trabeculae** (sing. **trabecula**), with many large spaces between them, filled by **red marrow** (see **bone marrow**).

Compact bone

A type of **bone** that forms the outer layer of all bones. It has far fewer spaces than **spongy bone**, and is laid in concentric layers (**lamellae**, sing. **lamella**) around channels called **Haversian canals**. These link with a complex system of tiny canals carrying blood vessels and nerves to the osteocytes.

Bone or osseous tissue

A type of tough connective tissue, made hard and resilient by large deposits of phosphorus and calcium compounds. The living bone cells, or **osteocytes**, are held in tiny spaces (**lacunae**, sing. **lacuna**) within this non-living material. There are two types of bone: **spongy** and **compact bone**.

Bone marrow

Two types of soft tissue. **Red marrow**, found in **spongy bone** (see **bone**), is where all new red (and some white) blood cells are made. **Yellow marrow** is a fat store, found in hollow areas (**medullary** or **marrow cavities**) in long bone **shafts**.

Tendons or sinews

Bands of tough connective tissue joining muscles to bones. Each is a continuation of the membrane around the muscle, together with the outer membranes of its bundles of fibres.

Cartilage or gristle

A tough connective tissue. In some joints (**cartilaginous joints**) it is the main cushion between the bones (e.g. **vertebrae***). In joints with **synovial sacs**, it covers the ends of the bones and is called **articular cartilage**. The end of the nose and the outer parts of the ears are made of cartilage. Young skeletons are made of cartilage, though these slowly turn to bone as minerals build up (a process called **ossification** or **osteogenesis**).

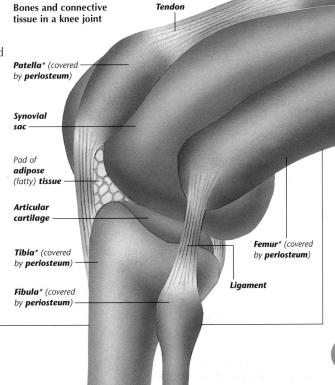

Bones and connective tissue in a knee joint

Tendon

Patella* (covered by **periosteum**)

Synovial sac

Pad of **adipose** (fatty) **tissue**

Articular cartilage

Tibia* (covered by **periosteum**)

Fibula* (covered by **periosteum**)

Femur* (covered by **periosteum**)

Ligament

MUSCLES

Muscles are areas of special elastic tissue (**muscle**) found all over the body. They may be either **voluntary muscles** (able to be controlled by conscious action) or **involuntary muscles** (not under conscious control). The main types of muscles are listed on this page.

Antagonistic pairs or opposing pairs

The pairs into which almost all muscles are arranged. The members of each pair produce opposite effects. In any given movement, the muscle which contracts to cause the movement is the **agonist** or **prime mover**. The one which relaxes at the same time is the **antagonist**.

Types of muscles

Skeletal muscles

All the muscles attached to the bones of the skeleton, which contract together or in sequence to move all the body parts. They are all **voluntary muscles** (see introduction) and are made of **striated muscle** tissue (see opposite). Some are named according to their position, shape or size, others are named after the movement they cause, e.g. **flexors** cause **flexion** (the bending of a limb at a joint), **extensors** straighten a limb.

Cardiac muscle

The muscle which makes up almost all of the wall of the heart. It is an **involuntary muscle** (see introduction) and is made of **cardiac muscle** tissue (see opposite).

Visceral muscles

The muscles in the walls of many internal organs, e.g. the intestines and blood vessels. They are all **involuntary muscles** (see introduction) and consist of **smooth muscle** tissue (see opposite).

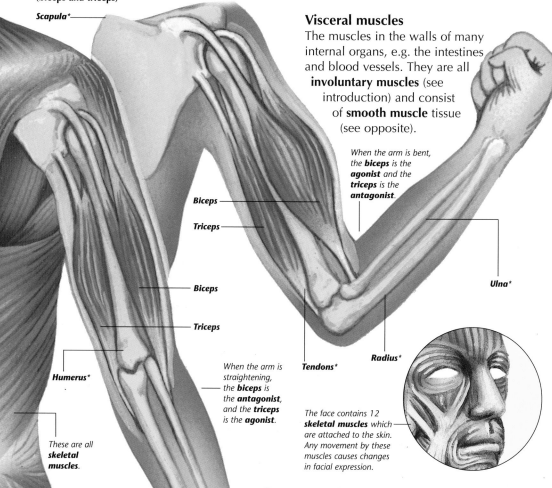

Example of an antagonistic pair (biceps and triceps)

*Scapula**

Biceps

Triceps

Biceps

Triceps

*Humerus**

These are all **skeletal muscles**.

*When the arm is straightening, the **biceps** is the **antagonist**, and the **triceps** is the **agonist**.*

*Tendons**

*When the arm is bent, the **biceps** is the **agonist** and the **triceps** is the **antagonist**.*

*Ulna**

*Radius**

The face contains 12 **skeletal muscles** which are attached to the skin. Any movement by these muscles causes changes in facial expression.

* **Humerus, Radius, Scapula**, 50; **Tendons**, 53; **Ulna**, 50.

The structure of muscle tissue

The different muscles in the body are made up of different kinds of muscle tissue (groups of cells of different types). The tissue has many blood vessels, bringing food matter to be broken down for energy, and nerves, which stimulate the muscles to act.

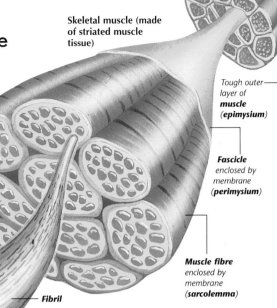

Skeletal muscle (made of striated muscle tissue)

Tough outer layer of **muscle** (**epimysium**)

Fascicle enclosed by membrane (**perimysium**)

Muscle fibre enclosed by membrane (**sarcolemma**)

Striated or striped muscle

The type of muscle tissue which makes up **skeletal muscles**. It consists of long cells called **muscle fibres**, grouped together in bundles called **fascicles**. Each fibre has a striped (**striated**) appearance under a microscope and is made of many smaller cylinders, called **fibrils** or **myofibrils**, which are the parts that contract when a fibre is stimulated by a nerve. The fibrils consist of interlocking **filaments**, or **myofilaments**, of two different types of protein – **actin** (thin filaments) and **myosin** (thicker filaments). These filaments slide past each other as a muscle contracts.

Filaments

— *Fibril*

(Relaxed) —

(Contracting)—

— **Myosin filaments**

— **Actin filaments**

Filaments slide past each other.

Cardiac muscle

A type of **striated muscle** tissue, making up the **cardiac muscle** of the heart. Its constant rhythmical contractions are caused by stimulations from areas of the tissue which produce electrical impulses. Any nervous impulses just increase or decrease this heart rate.

Smooth muscle or visceral muscle

The type of muscle tissue which makes up the **visceral muscles**. It consists of short, spindle-shaped cells. The way it contracts is not yet fully understood, but it contains **actin** and **myosin**, and is stimulated by nerves.

Nervous stimulation

Most muscles are stimulated to move by impulses from nerves running through the body. For more about this, see pages 80-81.

Muscle spindle

A group of **muscle fibres** (see **striated muscle**) which has the end fibres of a sensory nerve cell (**sensory neuron***) wrapped around it. The end fibres are part of one main fibre (**dendron***). When the muscle stretches they are stimulated to send impulses to the brain, "telling" it about the new state of tension. The brain can then work out the changes needed for any further action.

Motor end-plate

The point where the end fibres of an "instruction-carrying" nerve cell (**motor neuron***) meet a **muscle fibre** (see **striated muscle**). The end fibres are branches from one main fibre (**axon***). This carries nervous impulses which make the muscle contract. Each impulse is duplicated and sent down each end branch, hence the whole muscle receives a multiplication of each impulse.

— **Motor end-plate**

Axon* of **motor neuron***

* **Axon**, 76; **Dendron**, 76 (**Dendrites**);
Motor neurons, Sensory neurons, 77.

55

TEETH

The **teeth** or **dentes** (sing. **dens**) help to prepare food for digestion by cutting and grinding it up. Each tooth is set into the jaw, which has a soft tissue covering called **gum** (**gingiva**). During their lives, humans have two sets of teeth (**dentitions**) – a temporary set, or **deciduous dentition**, made up of 20 **deciduous teeth** (also called **milk** or **baby teeth**), and a later permanent dentition (32 **permanent teeth**).

Parts of a tooth

Crown
The exposed part of a tooth. It is covered by **enamel**. It is the part most subject to damage or tooth decay.

Root
The part of a tooth that is fixed in a socket in the jaw. **Incisors** and **canines** have one root, **premolars** have one or two and **molars** have two or three. Each root is held in place by the tough fibres of a **ligament*** called the **periodontal ligament**. The fibres are fixed to the jawbone at one end, and to the **cement** at the other. They act as shock absorbers.

Dentine or ivory
A yellow substance which forms the second layer inside a tooth. It is not as hard as **enamel** but, like it, has many of the same constituents as bone. It also contains **collagen*** fibres and strands of **cytoplasm***. These run out from the **pulp** cells in the **pulp cavity**.

Neck or cervix
The part of a tooth just below the surface, lying between the **crown** and the **root**.

Enamel
A substance similar to bone, though it is harder (the hardest substance in the body) and has no living cells. It consists of tightly-packed crystals of **apatite**, a mineral which contains calcium, phosphorus and fluorine.

Cement or cementum
A bone-like substance, similar to **enamel** but softer. It forms the thin surface layer of the **root** and is attached to the jaw by the **periodontal ligament** (see **root**).

Pulp cavity
The central area of a tooth, surrounded by **dentine**. It is filled with a soft tissue called **pulp**, which contains blood vessels and nerve fibre endings. These enter at the base of a **root** and run up to the cavity inside **root canals**. The blood vessels supply food and oxygen to the living tissue, and the nerve fibre endings are **pain receptors***.

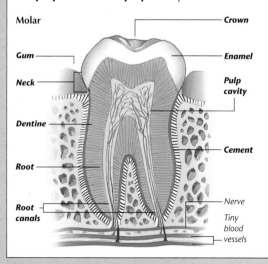

Molar — Crown
Gum — Enamel
Neck — Pulp cavity
Dentine
Cement
Root
Root canals — Nerve
Tiny blood vessels

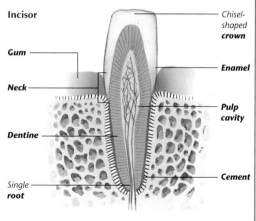

Incisor — Chisel-shaped crown
Gum — Enamel
Neck — Pulp cavity
Dentine
Single root — Cement

* **Collagen**, 52 (**Connective tissue**); **Cytoplasm**, 10; **Ligaments**, 52; **Pain receptors**, 83.

Types of teeth

Incisors

Sharp, chisel-shaped teeth, used for biting and cutting. Each has one root, and there are four in each jaw, set at the front of the mouth.

Canines or cuspids

Cone-shaped teeth (often called **eye** or **dog teeth**), used to tear food. Each has a sharp point (**cusp**) and one **root**. There are two in each jaw, one each side of the **incisors**. In many mammals, they are long and curved.

Premolars or bicuspids

Blunt, broad teeth, used for crushing and grinding (found in the permanent set of teeth only). There are four in each jaw, two behind each **canine**. Each has two sharp ridges (**cusps**) and one **root**, except the upper first premolars, which have two.

Molars

Blunt, broad teeth, similar to **premolars** but with a larger surface area. They are also used for crushing and grinding, and each has four surface points (**cusps**). Lower molars have two **roots** each, and upper ones have three. There are six molars in each jaw, three behind each pair of **premolars**, and the third ones (at the back) are known as **wisdom teeth**.

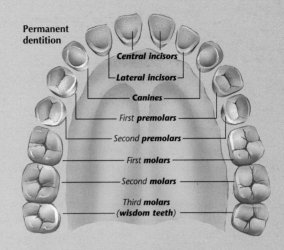

Permanent dentition

Central incisors
Lateral incisors
Canines
First **premolars**
Second **premolars**
First **molars**
Second **molars**
Third **molars**
(**wisdom teeth**)

 Incisors (replace eight temporary incisors).

 Premolars (replace eight temporary molars).

 Canines (replace four temporary canines).

 Molars (appear behind **premolars** and do not replace any deciduous teeth).

Wisdom teeth

Four **molars** (the third ones in line), lying at the end points of the jaws. They appear last of all, when a person is fully mature (hence their name). Often there is no room for them to come through and they get stuck in the jawbone, or **impacted**. A few people never develop wisdom teeth.

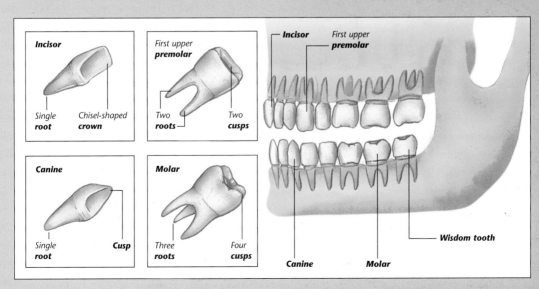

Incisor
Single **root**
Chisel-shaped **crown**

First upper premolar
Two **roots**
Two **cusps**

Canine
Single **root**
Cusp

Molar
Three **roots**
Four **cusps**

Incisor
First upper premolar
Wisdom tooth
Canine
Molar

BLOOD

Blood is a vital body fluid, consisting of **plasma**, **platelets** and **red** and **white blood cells**. An adult human has about 5.5 litres (9.5 pints), which travel around in the **circulatory system*** – a system of tubes called **blood vessels**. The blood distributes heat and carries many important substances in its plasma. Old, dying blood cells are constantly being replaced by new ones in a process called **haemopoiesis**.

Red blood cells

Blood constituents

Plasma
The pale liquid (about 90% water) which contains the blood cells. It carries dissolved food for the body cells, waste matter and carbon dioxide secreted by them, **antibodies** to combat infection, and **enzymes*** and **hormones*** which control body processes.

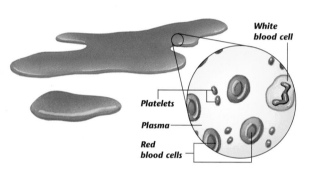

White blood cell

Platelets

Plasma

Red blood cells

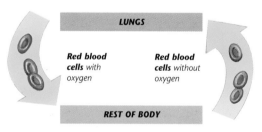

LUNGS

Red blood cells *with* oxygen

Red blood cells *without* oxygen

REST OF BODY

Platelets or thrombocytes
Very small, disc-shaped bodies with no **nuclei***, made in the **bone marrow***. They gather particularly at an injured area, where they are important in the **clotting** of blood.

White blood cells
Also called **white corpuscles** or **leucocytes**. Large, opaque blood cells, important in body defence. There are several types. **Lymphocytes**, for example, are made in **lymphoid tissue*** and are found in the **lymphatic system*** as well as blood. They make **antibodies**. Other white cells – **monocytes** – are made in **bone marrow***. They "swallow up" foreign bodies, e.g. bacteria, in a process called **phagocytosis***. Many of them (called **macrophages**) leave the blood vessels. They either travel around (**wandering macrophages**) or become fixed in an organ, e.g. a **lymph node*** (**fixed macrophages**).

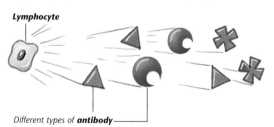

Lymphocyte

*Different types of **antibody***

Red blood cells
Also called **red corpuscles** or **erythrocytes**. Red, disc-shaped cells with no **nuclei***. They are made in the **bone marrow*** and contain **haemoglobin** (an iron compound which gives blood a dark red colour). This combines with oxygen in the lungs to form **oxyhaemoglobin**, and the blood becomes bright red. The red cells pass the oxygen to the body cells (by **diffusion***) and then return to the lungs with haemoglobin.

Monocyte

Bacterium

Pseudopodium*

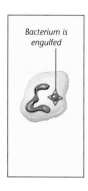

Bacterium is engulfed

58 ***Bone marrow**, 53; **Circulatory system**, 60; **Diffusion**, 101; **Enzymes**, 105; **Hormones**, 108; **Lymphatic system**, **Lymph nodes**, 65; **Lymphoid tissue**, 65 (**Lymphoid organs**); **Nucleus**, 10; **Phagocytosis**, 40 (**Pseudopodium**).

ABO blood groups

The main way of classifying blood. People with group A blood have A **antigen** on their **red blood cells** (and anti-B **antibodies**) and those with group B have B antigen (and anti-A antibodies). People with group AB have both antigens (and neither type of antibody) and those with group O have neither antigen (and both types of antibody).

Rhesus factor or Rh factor

A second way of classifying blood (as well as by **ABO blood group**). People whose **red blood cells** bear the **Rhesus antigen** are said to be **Rhesus positive**. Those without this antigen are **Rhesus negative**. Their blood does not normally contain anti-rhesus **antibodies**, but these would be produced if rhesus positive blood were to enter the body.

Body defence

Antibodies

"Defence" proteins in body fluids, e.g. **plasma**. They are made by **lymphocytes** (see **white blood cells**) to combat foreign **antigens** in the body. A different antibody is made for each antigen, and they act in different ways. **Anti-toxins** neutralize toxins (poisons). Each one joins with a toxin molecule, making an **antigen-antibody complex**. **Agglutinins** stick together the bacteria or viruses, and **lysins** kill them by dissolving their outer membranes.

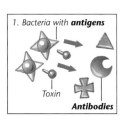

1. Bacteria with **antigens**

Toxin
Antibodies

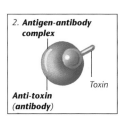

2. **Antigen-antibody complex**

Toxin
Anti-toxin (antibody)

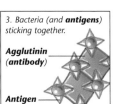

3. Bacteria (and **antigens**) sticking together.

Agglutinin (antibody)

Antigen

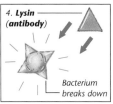

4. **Lysin (antibody)**

Bacterium breaks down

Antigens

Substances, mostly proteins, which cause the production of **antibodies** to combat them and any infection they may cause. They may form part of bacteria or viruses which enter the body, or they may be toxins (poisons) released by them. **ABO blood group** antigens and antibodies (see above) are present in the body from birth, ready to combat foreign blood group antigens.

Clotting or coagulation

The thickening of blood into a mass (**clot**) at the site of a wound. First, disintegrating **platelets** and damaged cells release a chemical called **thromboplastin**. This causes **prothrombin** (a **plasma** protein) to turn into **thrombin** (an **enzyme***). The thrombin then causes **fibrinogen** (another plasma protein) to harden into **fibrin**, a fibrous substance. A network of its fibres makes up the jelly-like clot.

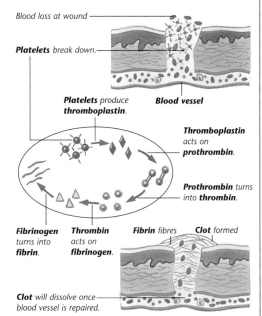

Blood loss at wound

Platelets break down.

Platelets produce **thromboplastin**.

Blood vessel

Thromboplastin acts on **prothrombin**.

Prothrombin turns into **thrombin**.

Fibrinogen turns into **fibrin**.

Thrombin acts on **fibrinogen**.

Fibrin fibres

Clot formed

Clot will dissolve once blood vessel is repaired.

Serum

A yellowy liquid consisting of the parts of the blood left after **clotting**. It contains many **antibodies** (produced to combat infections). When injected into other people, it can give temporary immunity to the infections.

*Enzymes, 105.

59

THE CIRCULATORY SYSTEM

The **circulatory** or **vascular system** is a network of blood-filled tubes, or **blood vessels**, of which there are three main types – **arteries**, **veins** and **capillaries**. A thin tissue layer called the **endothelium** lines arteries and veins, and is the only layer of capillary walls. Blood is kept flowing one way by the pumping of the heart, by muscles in artery and vein walls and by a decrease in pressure through the system (liquids flow from high to low pressure areas).

Passage of main substances in the circulatory system

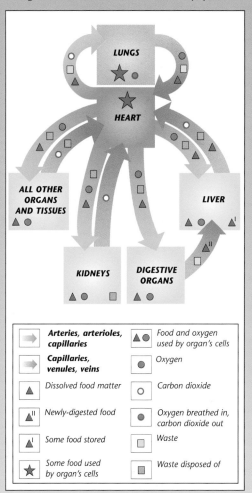

LUNGS

HEART

ALL OTHER ORGANS AND TISSUES

LIVER

KIDNEYS

DIGESTIVE ORGANS

→	Arteries, arterioles, capillaries	▲●	Food and oxygen used by organ's cells
→	Capillaries, venules, veins	●	Oxygen
▲	Dissolved food matter	○	Carbon dioxide
▲ᴵᴵ	Newly-digested food	◉	Oxygen breathed in, carbon dioxide out
▲ᴵ	Some food stored	□	Waste
★	Some food used by organ's cells	▣	Waste disposed of

Arteries

Wide, thick-walled blood vessels, making up the **arterial system** and carrying blood away from the heart. Smaller arteries (**arterioles**) branch off the main ones, and **capillaries** branch off the arterioles. Except in the **pulmonary arteries***, the blood contains oxygen (which makes it bright red). In all arteries it also carries dissolved food and waste, brought into the heart by **veins**, and there transferred to the arteries. These carry the food to the cells (via arterioles and capillaries) and the waste to the kidneys.

Artery

Outer layer —
Smooth muscle* —
— Elastic fibrous tissue
— Endothelium

Veins

Wide, thick-walled blood vessels, making up the **venous system** and carrying blood back to the heart. They contain valves to stop blood flowing backwards due to gravity, and are formed from merging **venules** (small veins). These are formed in turn from merging **capillaries**. The blood contains carbon dioxide (except in the **pulmonary veins***) and waste matter, both picked up from body cells by the capillaries. The blood in the veins leading from the digestive system and liver also carries dissolved food. This is transferred to the arteries in the heart.

Vein

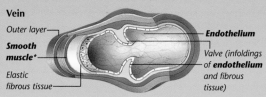

Outer layer —
Smooth muscle* —
Elastic fibrous tissue —
— Endothelium
Valve (infoldings of **endothelium** and fibrous tissue)

Capillaries

Narrow, thin-walled blood vessels, branching off **arterioles** (see **arteries**) to form a complex network. Oxygen and dissolved food pass out through their walls to the body cells, and carbon dioxide and waste pass in (see **tissue fluid**, page 64). The capillaries of the digestive organs and liver also pick up food. Capillaries finally join up again to form small veins (**venules**).

Capillary

Single layer (**endothelium**)

***Hepatic portal vein**, 69 (**Liver**); **Pulmonary arteries**, 63 (**Pulmonary trunk**); **Pulmonary veins**, 63; **Smooth muscle**, 55.

The main arteries and veins

Vascular means "composed of or containing conducting vessels". In the case of animals, it means having a blood supply.

Avascular means "containing no conducting vessels". In the case of animals, it means having no blood supply.

The main **blood vessels** of the head, heart and lungs are named on page 62.

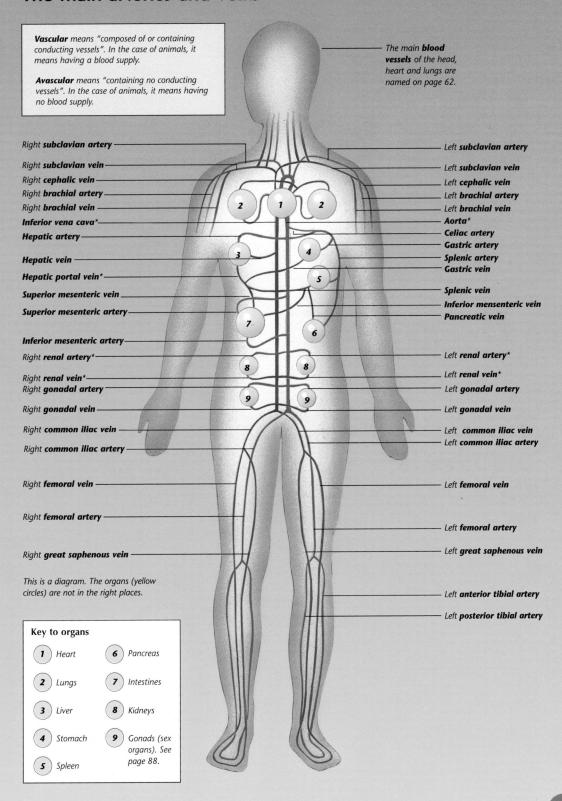

Right **subclavian artery**

Right **subclavian vein**
Right **cephalic vein**
Right **brachial artery**
Right **brachial vein**
Inferior vena cava*
Hepatic artery

Hepatic vein

Hepatic portal vein*

Superior **mesenteric vein**

Superior **mesenteric artery**

Inferior mesenteric artery
Right **renal artery***

Right **renal vein***
Right **gonadal artery**

Right **gonadal vein**

Right **common iliac vein**

Right **common iliac artery**

Right **femoral vein**

Right **femoral artery**

Right **great saphenous vein**

Left **subclavian artery**

Left **subclavian vein**
Left **cephalic vein**
Left **brachial artery**
Left **brachial vein**
Aorta*
Celiac artery
Gastric artery
Splenic artery
Gastric vein

Splenic vein
Inferior mensenteric vein
Pancreatic vein

Left **renal artery***

Left **renal vein***
Left **gonadal artery**

Left **gonadal vein**

Left **common iliac vein**
Left **common iliac artery**

Left **femoral vein**

Left **femoral artery**

Left **great saphenous vein**

Left **anterior tibial artery**

Left **posterior tibial artery**

This is a diagram. The organs (yellow circles) are not in the right places.

Key to organs

1 Heart
2 Lungs
3 Liver
4 Stomach
5 Spleen
6 Pancreas
7 Intestines
8 Kidneys
9 Gonads (sex organs). See page 88.

* **Aorta**, 63; **Hepatic portal vein**, 69 (**Liver**); **Inferior vena cava**, 63; **Renal arteries**, **Renal veins**, 72 (**Kidneys**).

THE HEART

The **heart** is a muscular organ which pumps blood around the blood vessels. (The heart and blood vessels together are the **cardiovascular system.**) It is surrounded by the **pericardial sac.** This consists of an outer membrane (the **pericardium**) and the cavity (**pericardial cavity**) between it and the heart. This cavity is filled with a cushioning fluid (**pericardial fluid**). The heart has four chambers – two **atria** and two **ventricles**, all lined by a thin tissue layer called the **endocardium.**

Position of **heart**

The chambers of the heart

Atria (sing. atrium) or auricles
The two upper chambers. The left atrium receives **oxygenated** blood (blood with fresh oxygen – see also **haemoglobin***) from the lungs via the **pulmonary veins**. The right atrium receives **deoxygenated** blood from the rest of the body via the **superior** and **inferior vena cavae.** This is blood whose oxygen has been used by the cells and replaced by carbon dioxide.

Ventricles
The two lower chambers. The left ventricle receives blood from the left **atrium** and pumps it into the **aorta**. The right ventricle receives blood from the right atrium and pumps it via the **pulmonary trunk** to the lungs.

Key	
➡	**Oxygenated** blood
➡	**Deoxygenated** blood

Cardiac means "of or near the heart". **Pulmonary** means "of the lungs".

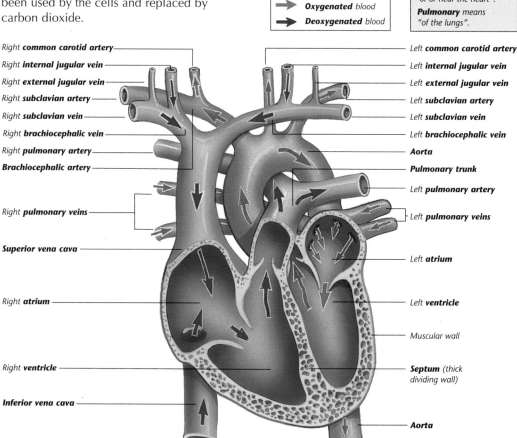

Right **common carotid artery**
Right **internal jugular vein**
Right **external jugular vein**
Right **subclavian artery**
Right **subclavian vein**
Right **brachiocephalic vein**
Right **pulmonary artery**
Brachiocephalic artery
Right **pulmonary veins**
Superior vena cava
Right **atrium**
Right **ventricle**
Inferior vena cava

Left **common carotid artery**
Left **internal jugular vein**
Left **external jugular vein**
Left **subclavian artery**
Left **subclavian vein**
Left **brachiocephalic vein**
Aorta
Pulmonary trunk
Left **pulmonary artery**
Left **pulmonary veins**
Left **atrium**
Left **ventricle**
Muscular wall
Septum (thick dividing wall)
Aorta

* **Haemoglobin**, 58 (**Red blood cells**).

The main arteries and veins

Aorta
The largest **artery*** in the body. It carries blood with fresh oxygen out of the left **ventricle** to begin its journey round the body.

Pulmonary trunk
The **artery*** which carries blood needing fresh oxygen out of the right **ventricle**. After leaving the heart, it splits into the right and left **pulmonary arteries**, one going to each lung.

Superior vena cava
One of the two main **veins***. It carries blood needing fresh oxygen from the upper body to the right **atrium**. All the upper body veins merge into it.

Inferior vena cava
One of the two main **veins***, carrying blood needing fresh oxygen from the lower body to the right **atrium**. All the lower body veins merge into it.

Pulmonary veins
Four **veins*** which carry blood with fresh oxygen to the left **atrium**. Two right pulmonary veins come from the right lung, and two left pulmonary veins come from the left lung.

Semilunar valves
Two valves, so called because they have crescent-shaped flaps. One is the **aortic valve** between the left **ventricle** and the **aorta**. The other is the **pulmonary valve** between the right ventricle and the **pulmonary trunk**.

Closed flaps of **pulmonary valve**

Open flaps of **aortic valve**

Atrioventricular valves or AV valves
Two valves, each between an **atrium** and its corresponding **ventricle**. The left AV valve, or **mitral valve**, is a **bicuspid valve**, i.e. it has two movable flaps, or **cusps**. The right AV valve is a **tricuspid valve**, i.e. it has three cusps.

Closed **cusps** *of* **left AV valve**

Open **cusps** *of* **right AV valve**

The cardiac cycle

The **cardiac cycle** is the series of events which make up one complete pumping action of the heart, and which can be heard as the heartbeat (about 70 times a minute). First, both **atria** contract and pump blood into their respective **ventricles**, which relax to receive it. Then the atria relax and take in blood, and the ventricles contract to pump it out. The relaxing phase of a chamber is its **diastole phase**; the contracting phase is its **systole phase**. There is a short pause after the systole phase of the ventricles, during which all chambers are in diastole phase (relaxing). The different **valves** which open and close during the cycle are defined below left.

Cardiac cycle

1. Atria *in* **systole phase**, *ventricles in* **diastole phase**.

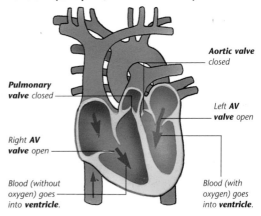

Aortic valve closed

Pulmonary valve closed

*Left **AV** valve* open

*Right **AV** valve* open

Blood (without oxygen) goes into **ventricle**.

Blood (with oxygen) goes into **ventricle**.

2. Atria *in* **diastole phase**, *ventricles in* **systole phase**.

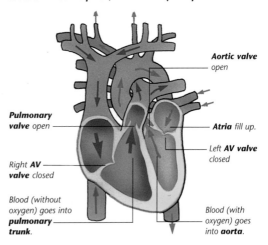

Aortic valve open

Pulmonary valve open

Atria fill up.

*Left **AV** valve* closed

*Right **AV** valve* closed

Blood (without oxygen) goes into **pulmonary trunk**.

Blood (with oxygen) goes into **aorta**.

TISSUE FLUID AND THE LYMPHATIC SYSTEM

The smallest blood vessels, called **capillaries***, are those in the most direct contact with the individual cells of the body, but even they do not touch the cells. The food and oxygen they carry finally reach the cells in **tissue fluid**, a substance which forms the link between the **circulatory system*** and the body's drainage system, known as the **lymphatic system**.

Tissue fluid

Also called **intercellular** or **interstitial fluid**. A fluid which surrounds the body cells. It seeps out from the blood through the walls of **capillaries*** (mainly at their high-pressure ends, after they have branched from **arterioles***) and is essentially **plasma***, though with fewer proteins. It carries oxygen and dissolved food to the body cells, and carbon dioxide and waste matter away from them. These latter substances enter the capillaries (mainly at their low-pressure ends, before they form **venules***).

The protein molecules not needed by the cells are too large to re-enter the capillaries. They pass, with some of the waste, into the **lymph capillaries** (see **lymph vessels**), whose walls are more easily penetrated.

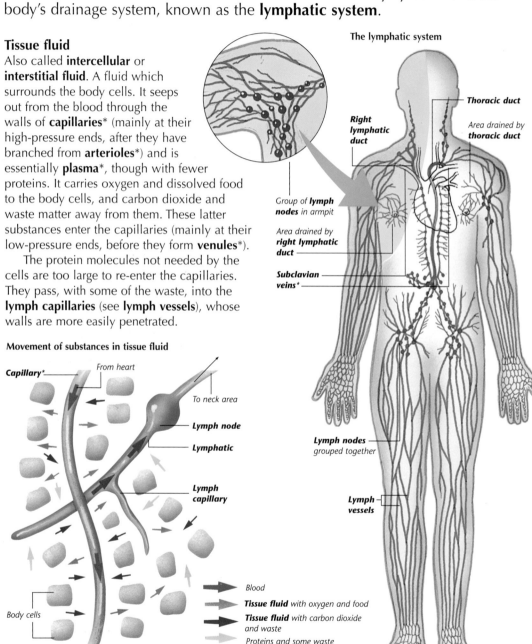

The lymphatic system

Thoracic duct

Right lymphatic duct

Area drained by **thoracic duct**

Group of **lymph nodes** in armpit

Area drained by **right lymphatic duct**

Subclavian veins*

Lymph nodes grouped together

Lymph vessels

Movement of substances in tissue fluid

Capillary*

From heart

To neck area

Lymph node

Lymphatic

Lymph capillary

Body cells

To heart

Blood

Tissue fluid with oxygen and food

Tissue fluid with carbon dioxide and waste

Proteins and some waste

Lymph

* **Arterioles**, 60 (**Arteries**); **Capillaries**, **Circulatory system**, 60;
Plasma, 58; **Subclavian veins**, 61; **Venules**, 60 (**Veins**).

Lymphatic system

A system of tubes (**lymph vessels**) and small organs (**lymphoid organs**), important in the recycling of body fluids and in the fight against disease. The lymph vessels carry the liquid **lymph** around the body and empty it back into the **veins***, and the lymphoid organs are the source of disease-fighting cells.

Lymph

The liquid in **lymph vessels**. It contains **lymphocytes** (see **lymphoid organs**), some substances picked up from **tissue fluid** (especially proteins such as **hormones*** and **enzymes***) and also fat particles (see **lymph vessels**).

Lymph vessels or lymphatic vessels

Blind-ended tubes carrying **lymph** from all body areas towards the neck, where it is emptied back into the blood. They are lined with **endothelium***, and have valves to stop the lymph from being pulled back by gravity.

The thinnest lymph vessesls are **lymph capillaries**, and include the important **lacteals***, which pick up fat particles (too large to enter the bloodstream directly). The capillaries join to form larger vessels called **lymphatics**, which finally unite to form two tubes – the **right lymphatic duct** (emptying into the right **subclavian vein***) and the **thoracic duct** (emptying into the left **subclavian vein***).

Lymphoid organs

The **lymphoid organs**, or **lymphatic organs**, are bodies connected to the **lymphatic system**. They are all made of the same type of tissue (**lymphoid** or **lymphatic tissue**) and they all produce **lymphocytes*** – disease-fighting white blood cells.

Lymph nodes or lymph glands

Small lymphoid organs found along the course of **lymph vessels**, often in groups, e.g. in the armpits. They are the main sites of **lymphocyte** production (see above) and also contain a filter system which traps bacteria and foreign bodies. These are then engulfed by white blood cells (**fixed macrophages***).

Spleen

The largest lymphoid organ, found just below the **diaphragm*** on the left side of the body. It holds an emergency store of red blood cells and also contains white blood cells (**fixed macrophages***) which destroy foreign bodies, e.g. bacteria, and old blood cells.

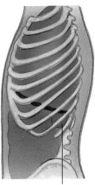

*Position of **spleen***

Thymus gland

A lymphoid organ in the upper part of the chest. It is fairly large in children, reaches its maximum size at **puberty*** and then undergoes **atrophy**, i.e. it wastes away.

Tonsils

Four lymphoid organs: one **pharyngeal tonsil** (the **adenoids**) at the back of the nose, one **lingual tonsil** at the base of the tongue and two **palatine tonsils** at the back of the mouth.

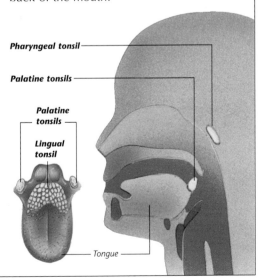

Pharyngeal tonsil

Palatine tonsils

Palatine tonsils

Lingual tonsil

Tongue

*** Diaphragm**, 70; **Endothelium**, 60; **Enzymes**, 105; **Fixed macrophages**, 58 (**White blood cells**); **Hormones**, 108; **Lacteals**, 67 (**Small intestine**); **Lymphocytes**, 58 (**White blood cells**); **Puberty**, 90; **Subclavian veins**, 61; **Veins**, 60.

THE DIGESTIVE SYSTEM

After food is taken in, or **ingested**, it passes through the **digestive system**, gradually being broken down into simple soluble substances by a process called **digestion** (see also pages 110-111). The simple substances are absorbed into the blood vessels around the system and transported to the body cells. Here they are used to provide energy and build new tissue. For more about all these different processes, see pages 102-107. The main parts of the digestive system are listed on these two pages. The pancreas and liver (see page 69) also play a vital part in digestion, forming the two main **digestive glands*** (producing **digestive juices***).

Position of
digestive system

Alimentary canal

Also called the **alimentary tract**, **gastrointestinal (GI) tract**, **enteric canal** or the **gut**. A collective term for all the parts of the digestive system. It is a long tube running from the mouth to the **anus** (see **large intestine**). Most of its parts are in the lower body, or **abdomen**, inside the main body cavity, or **perivisceral cavity***. They are held in place by **mesenteries** – infoldings of the cavity lining (the **peritoneum**).

Pharynx

A cavity at the back of the mouth, where the mouth cavity (**oral or buccal cavity**) and the **nasal cavities*** meet. When food is swallowed, the **soft palate** (a tissue flap at the back of the mouth) closes the nasal cavities and the **epiglottis*** closes the **trachea***.

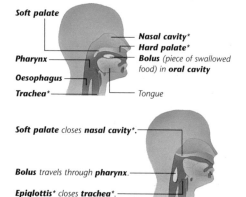

Soft palate

Nasal cavity*
Hard palate*
Pharynx —
Bolus (piece of swallowed food) in **oral cavity**
Oesophagus
Trachea* —
Tongue

Soft palate closes **nasal cavity***.

Bolus travels through **pharynx**.

Epiglottis* closes **trachea***.

Oesophagus or gullet

The tube down which food travels to the **stomach**. A piece of swallowed food is a **bolus**.

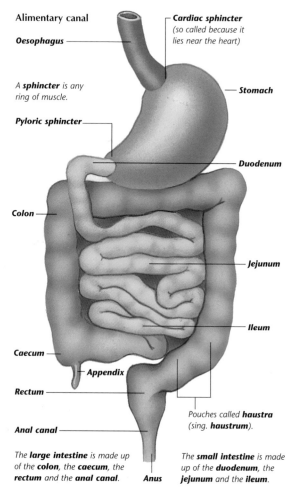

Alimentary canal

Oesophagus

A **sphincter** is any ring of muscle.

Pyloric sphincter

Colon

Caecum

Rectum

Anal canal

The **large intestine** is made up of the **colon**, the **caecum**, the **rectum** and the **anal canal**.

Cardiac sphincter (so called because it lies near the heart)

Stomach

Duodenum

Jejunum

Ileum

Appendix

Pouches called **haustra** (sing. **haustrum**).

Anus

The **small intestine** is made up of the **duodenum**, the **jejunum** and the **ileum**.

*Digestive juices, 68 (**Digestive glands**); **Epiglottis**, 70; **Hard palate**, 79; **Nasal cavities**, 79 (**Nose**); **Perivisceral cavity**, 37; **Trachea**, 70.

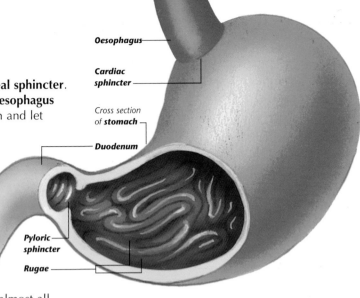

Cardiac sphincter

Also called the **gastroesophageal sphincter**. A muscular ring between the **oesophagus** and **stomach**. It relaxes to open and let food through.

Stomach

A large sac in which the early stages of digestion occur. Its lining has many folds (**rugae**, sing. **ruga**) which flatten out to let it expand. Some substances, e.g. water, pass through its wall into nearby blood vessels, but almost all the semi-digested food (**chyme**) goes into the **small intestine (duodenum)**.

Pyloric sphincter

Also called the **pyloric valve** or **pylorus**. A muscular ring between the **stomach** and the **small intestine**. It relaxes to let food through only after certain digestive changes have occurred.

Small intestine

The main site of digestion, a coiled tube with three parts – the **duodenum**, **jejunum** and **ileum**. Many tiny "fingers" called **villi** (sing. **villus**) project inwards from its lining. Each contains **capillaries*** (tiny blood vessels) into which most of the food is absorbed, and a **lymph vessel*** called a **lacteal**, which absorbs recombined fat particles (see **fats**, page 102). The remaining semi-liquid waste mixture passes into the **large intestine**.

Large intestine

A thick tube receiving waste from the **small intestine**. It consists of the **caecum*** (a redundant sac), **colon**, **rectum** and **anal canal**. The colon contains bacteria, which break down any remaining food and make some important vitamins. Most of the water in the waste passes through the colon walls into nearby blood vessels. This leaves a semi-solid mass (**faeces**), which is pushed out of the body (**defaecation**) via the rectum, anal canal and **anus** – a hole surrounded by a muscular ring (the **anal sphincter**).

Appendix

A small, blind-ended tube off the **caecum** (see **large intestine**). It is a **vestigial** organ, i.e. one which our ancestors needed, but is now defunct.

Mucous membrane or mucosa

A thin layer of tissue lining all digestive passages (also other passages, e.g. the air passages). It is a special type of **epithelium*** (a surface sheet of cells), containing many single-celled **exocrine glands***, called **mucous glands**. These secrete **mucus** – a lubricating fluid which, in the case of the digestive passages, also protects the passage walls against the action of **digestive juices***.

Peristalsis

The waves of contraction, produced by muscles in the walls of organs (especially digestive organs), which move substances along.

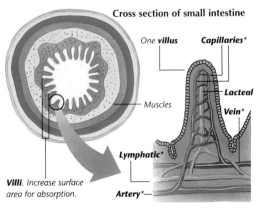

Cross section of small intestine

One *villus* Capillaries*

Muscles

Lacteal

Vein*

Lymphatic*

Villi. Increase surface area for absorption.

Artery*

* **Arteries**, 60; **Caecum**, 43; **Capillaries**, 60; **Digestive juices**, 68 (**Digestive glands**); **Epithelium**, 82 (**Epidermis**); **Exocrine glands**, 68; **Lymphatic**, 65 (**Lymph vessels**); **Veins**, 60.

67

GLANDS

Glands are special organs (or sometimes groups of cells or single cells) which produce and secrete a variety of substances vital to life. There are two types of human gland – **exocrine** and **endocrine**.

Exocrine glands

Exocrine glands are glands which secrete substances through tubes, or **ducts**, onto a surface or into a cavity. Most body glands are exocrine, e.g. **sweat glands*** and **digestive glands**.

Digestive glands

Exocrine glands which secrete fluids called **digestive juices** into the digestive organs. The juices contain **enzymes*** which cause the breakdown of food (see chart, pages 110-111). Many of the glands are tiny, and set into the walls of the digestive organs, e.g. **gastric glands** in the stomach and **intestinal glands** (or **crypts of Lieberkühn**) in the small intestine. Others are larger and lie more freely, e.g. **salivary glands**. The largest are the **pancreas** and **liver**.

Salivary glands (secrete **saliva*** into mouth)

Only one side is shown – the three glands are duplicated on the other side.

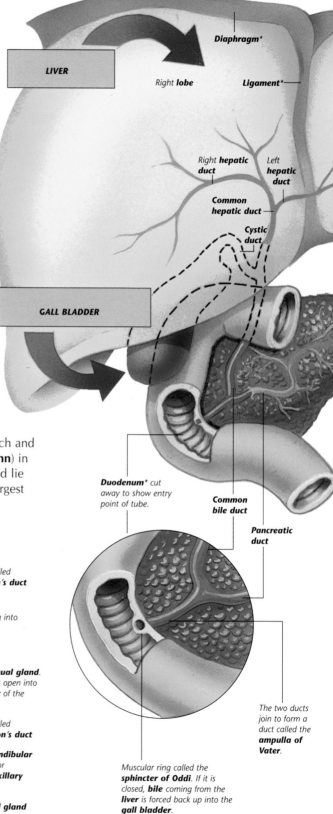

LIVER

Diaphragm*

Right **lobe**

Ligament*

Right **hepatic duct**

Left **hepatic duct**

Common hepatic duct

Cystic duct

GALL BLADDER

Duodenum* cut away to show entry point of tube.

Common bile duct

Pancreatic duct

The two ducts join to form a duct called the **ampulla of Vater**.

Muscular ring called the **sphincter of Oddi**. If it is closed, **bile** coming from the **liver** is forced back up into the **gall bladder**.

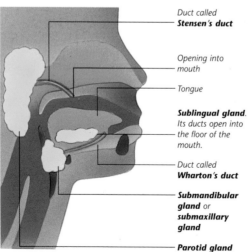

Duct called **Stensen's duct**

Opening into mouth

Tongue

Sublingual gland. Its ducts open into the floor of the mouth.

Duct called **Wharton's duct**

Submandibular gland or **submaxillary gland**

Parotid gland

***Diaphragm**, 70; **Duodenum**, 67 (**Small intestine**); **Enzymes**, 105; **Ligaments**, 52; **Saliva**, 110; **Sweat glands**, 83.

Liver

*Left **lobe***

The largest organ. One of its many roles is that of a **digestive gland**, secreting **bile** (see chart, pages 110-111) along the **common hepatic duct**. Another vital job is the conversion and storage of newly-digested food matter (see diagram, page 103), which it receives along the **hepatic portal vein** (see picture, page 61). In particular, it regulates the amount of glucose in the blood. It also destroys worn-out red blood cells, stores vitamins and iron, and makes important blood proteins.

PANCREAS

Pancreas

A large gland which is both a **digestive gland** and an **endocrine gland**. It produces **pancreatic juice** (see chart, pages 110-111), which it secretes along the **pancreatic duct**, or **duct of Wirsung**. It also contains groups of cells called the **islets of Langerhans**. These make up the endocrine parts of the organ, and produce the **hormones*** **insulin*** and **glucagon***.

Gall bladder

A sac which stores **bile** (made in the **liver**) in a concentrated form until it is needed (i.e. until there is food in the **duodenum***). Its lining has many folds (**rugae**, sing. **ruga**) which flatten out as it expands. When needed, the bile is squeezed along the **cystic duct** and the **common bile duct**.

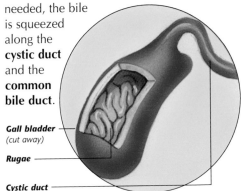

Gall bladder (cut away)

Rugae

Cystic duct

Endocrine glands

Endocrine or **ductless glands** are glands which secrete substances called **hormones** directly into the blood (i.e. blood vessels in the glands). For more about hormones, see pages 108-109. The glands may be separate bodies (e.g. those below) or cells inside organs, e.g. in the sex organs.

Pituitary gland

Also called the **pituitary body** or **hypophysis**. A gland at the base of the brain, directly influenced by the **hypothalamus*** (see also **hormones**, page 108) and made up of an **anterior** (front) **lobe** (**adenohypophysis**) and a **posterior** (back) **lobe** (**neurohypophysis**). Many of its hormones are **tropic hormones**, i.e. they stimulate other glands to secrete hormones. It makes **ACTH, TSH, STH, FSH, LH, lactogenic hormone, oxytocin** and **ADH**.

Thyroid gland

A large gland around the **larynx***. It produces **thyroxin** and **thyrocalcitonin**.

Parathyroid glands

Two pairs of small glands embedded in the **thyroid gland**. They produce **PTH**.

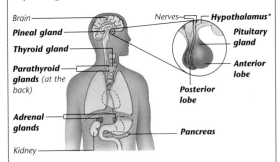

Brain

Pineal gland

Thyroid gland

Parathyroid glands (at the back)

Adrenal glands

Kidney

Nerves

Hypothalamus*

Pituitary gland

Anterior lobe

Posterior lobe

Pancreas

Adrenal glands or suprarenal glands

A pair of glands, one gland lying above each kidney. Each has an outer layer (**cortex**), producing **aldosterone, cortisone** and **hydrocortisone**, and an inner layer (**medulla**), producing **adrenalin** and **noradrenalin**.

Pineal gland

Also called the **pineal body**. A small gland at the front of the brain. Its role is not clear, but it is known to secrete **melatonin**, a hormone thought to influence **sex hormone*** production.

* **Duodenum**, 67 (**Small intestine**); **Glucagon, Hormones**, 108; **Hypothalamus**, 75; **Insulin**, 108; **Larynx**, 70; **Sex hormones**, 108.

THE RESPIRATORY SYSTEM

The term **respiration** covers three processes: **ventilation**, or breathing (taking in oxygen and expelling carbon dioxide), **external respiration** (the exchange of gases between the **lungs** and the blood – see also **red blood cells**, page 58) and **internal respiration** (food breakdown, using oxygen and producing carbon dioxide – see pages 106-107). Listed here are the component parts of the human **respiratory system**.

Position of
respiratory system

Trachea or windpipe
The main tube through which air passes on its way to and from the **lungs**.

Larynx
The "voice box" at the top of the **trachea**. It contains the **vocal cords** – two pieces of tissue folding inwards from the trachea lining and attached to plates of **cartilage***. The opening between the cords is called the **glottis**. During speech, muscles pull the cartilage plates (and hence the cords) together, and air passing out through the cords makes them vibrate, producing sounds.

Lungs
The two main breathing organs, inside which gases are exchanged. They contain many tubes (**bronchi** and **bronchioles**) and air sacs (**alveoli**).

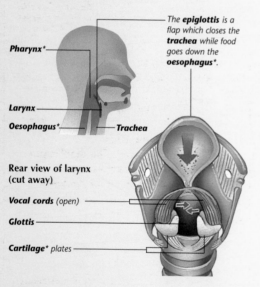

The **epiglottis** is a flap which closes the **trachea** while food goes down the **oesophagus***.

Pharynx*

Larynx

Oesophagus*

Trachea

Rear view of larynx (cut away)

Vocal cords (open)

Glottis

Cartilage* plates

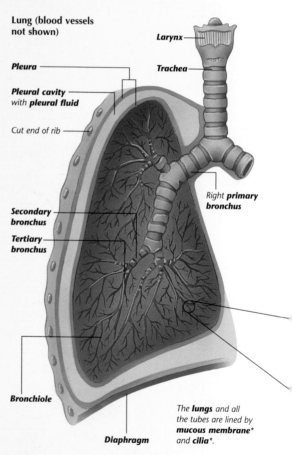

Lung (blood vessels not shown)

Larynx

Pleura

Trachea

Pleural cavity with **pleural fluid**

Cut end of rib

Right **primary bronchus**

Secondary bronchus

Tertiary bronchus

Bronchiole

Diaphragm

The **lungs** and all the tubes are lined by **mucous membrane*** and **cilia***.

Pleura (sing. pleuron) or pleural membrane
A layer of tissue surrounding each **lung** and lining the chest cavity (**thorax**). Between the pleura around a lung and the pleura lining the thorax there is a space (**pleural cavity**). This contains **pleural fluid**. The pleura and fluid-filled cavity make up a cushioning **pleural sac**.

Diaphragm or midriff
A sheet of muscular tissue which separates the chest from the lower body, or **abdomen**. At rest, it lies in an arched position, forced up by the abdomen wall below it.

* **Cartilage**, 53; **Cilia**, 40; **Mucous membrane**, 67; **Oesophagus**, **Pharynx**, 66.

Bronchi (sing. bronchus)

The main tubes into which the **trachea** divides. The first two branches are the right and left **primary bronchi**. Each carries air into a **lung** (via a hole called a **hilum**), alongside a **pulmonary artery*** bringing blood in. They then branch into **secondary bronchi**, **tertiary bronchi** and **bronchioles**, all accompanied by blood vessels, both branching from the pulmonary artery and merging to form **pulmonary veins*** (blood going out).

Bronchioles

The millions of tiny tubes in the **lungs**, all accompanied by blood vessels. They branch off **tertiary bronchi** (see **bronchi**) and have smaller branches called **terminal bronchioles**, each one ending in a cluster of **alveoli**.

Alveoli (sing. alveolus)

The millions of tiny sacs attached to **terminal bronchioles** (see **bronchioles**). They are surrounded by **capillaries*** (tiny blood vessels) whose blood is rich in carbon dioxide. This passes out through the capillary walls, and in through those of the alveoli (to be breathed out). The oxygen which has been breathed into the alveoli passes into the capillaries, which then begin to merge together (eventually forming **pulmonary veins***).

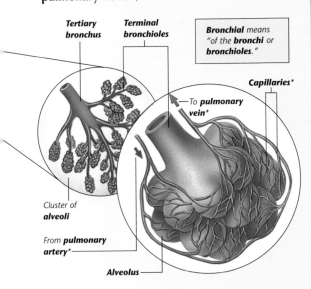

Tertiary bronchus

Terminal bronchioles

> **Bronchial** means "of the **bronchi** or **bronchioles**."

Capillaries*

To **pulmonary vein***

Cluster of **alveoli**

From **pulmonary artery***

Alveolus

Breathing

Breathing is made up of **inspiration** (breathing in) and **expiration** (breathing out). Both actions are normally automatic, controlled by nerves from the **respiratory centre** in the **medulla*** of the brain. This acts when it detects too high a level of carbon dioxide in the blood.

Inspiration or inhalation

The act of breathing in. The **diaphragm** contracts and flattens, lengthening the chest cavity. The muscles between the ribs (**intercostal muscles**) also contract, pulling the ribs up and outwards and widening the cavity. The overall expansion lowers the air pressure in the **lungs**, and air rushes in to fill them (i.e. to equalize internal and external pressure).

Inspiration

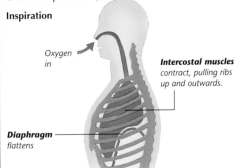

Oxygen in

Intercostal muscles contract, pulling ribs up and outwards.

Diaphragm flattens

Expiration or exhalation

The act of breathing out. The **diaphragm** and **intercostal muscles** (see **inspiration**) relax, and air is forced out of the **lungs** as the chest cavity becomes smaller.

Expiration

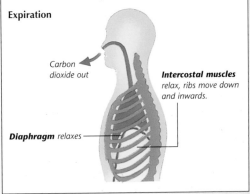

Carbon dioxide out

Intercostal muscles relax, ribs move down and inwards.

Diaphragm relaxes

* **Capillaries**, 60; **Medulla**, 75; **Pulmonary arteries**, 63 (**Pulmonary trunk**); **Pulmonary veins**, 63.

THE URINARY SYSTEM

The **urinary system** is the main system of body parts involved in **excretion**, which is the expulsion of unwanted substances. The parts are defined below. The lungs and skin are also involved in excretion (expelling carbon dioxide and sweat respectively).

Parts of the urinary system

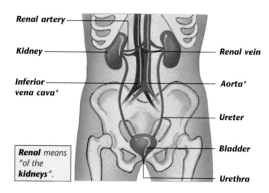

Renal artery
Kidney
Inferior vena cava*
Renal vein
Aorta*
Ureter
Bladder
Urethra

Renal means "of the kidneys".

Kidneys

Two organs at the back of the body, just below the ribs. They are the main organs of excretion, filtering out unwanted substances from the blood and regulating the level and contents of body fluids (see also **homeostasis**, page 107). Blood enters a kidney in a **renal artery** and leaves it in a **renal vein**.

Ureters

The two tubes which carry **urine** from the **kidneys** to the **bladder**.

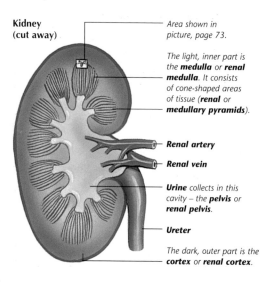

Kidney (cut away)

Area shown in picture, page 73.

The light, inner part is the **medulla** or **renal medulla**. It consists of cone-shaped areas of tissue (**renal** or **medullary pyramids**).

Renal artery

Renal vein

Urine collects in this cavity – the **pelvis** or **renal pelvis**.

Ureter

The dark, outer part is the **cortex** or **renal cortex**.

Bladder or urinary bladder

A sac which holds stored **urine**. Its lining has many folds (**rugae**, sing. **ruga**) which flatten out as it fills up, letting it expand. Two muscular rings – the **internal** and **external urinary sphincters** – control the opening from the bladder into the **urethra**. When the volume of urine gets to a certain level, nerves stimulate the internal sphincter to open, but the external sphincter is under conscious control (except in young children), and can be held closed for longer.

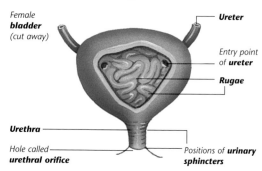

Female bladder (cut away)
Ureter
Entry point of **ureter**
Rugae
Urethra
Hole called **urethral orifice**
Positions of **urinary sphincters**

Urethra

The tube carrying **urine** from the **bladder** out of the body (in men, it also carries **sperm*** – see **penis**, page 88). The expulsion of urine is called **urination** or **micturition**.

Urea

A nitrogen-containing (**nitrogenous**) waste substance which is a product of the breakdown of excess **amino acids*** in the liver. It travels in the blood to the **kidneys**, together with smaller amounts of similar substances, e.g. creatinine.

Urine

The liquid which leaves the **kidneys**. Its main constituents are excess water, **urea** and minerals.

*Amino acids,102 (**Proteins**); Aorta, Inferior vena cava, 63; Sperm, 92 (**Gametes**).

Inside a kidney

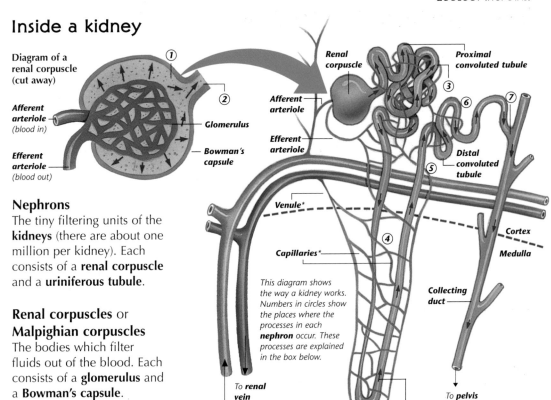

Diagram of a renal corpuscle (cut away)

Afferent arteriole (blood in)

Glomerulus

Efferent arteriole (blood out)

Bowman's capsule

Renal corpuscle

Proximal convoluted tubule

Afferent arteriole

Efferent arteriole

Distal convoluted tubule

Venule*

Cortex

Medulla

Capillaries*

This diagram shows the way a kidney works. Numbers in circles show the places where the processes in each **nephron** occur. These processes are explained in the box below.

Collecting duct

To **renal vein**

From **renal artery**

Loop of Henlé

To **pelvis**

Nephrons
The tiny filtering units of the **kidneys** (there are about one million per kidney). Each consists of a **renal corpuscle** and a **uriniferous tubule**.

Renal corpuscles or Malpighian corpuscles
The bodies which filter fluids out of the blood. Each consists of a **glomerulus** and a **Bowman's capsule**.

Glomerulus
A ball of coiled-up **capillaries*** (tiny blood vessels) at the centre of each **renal corpuscle**. The capillaries branch from an **arteriole*** entering the corpuscle (an **afferent arteriole**) and re-unite to leave the corpuscle as an **efferent arteriole**.

Bowman's capsule
The outer part of each **renal corpuscle**. It is a thin-walled sac around the **glomerulus**.

Uriniferous tubules or renal tubules
Long tubes, each one leading from a **Bowman's capsule**. Each has three main parts – the **proximal convoluted tubule**, the **loop of Henlé** and the **distal convoluted tubule** – and has many **capillaries*** (tiny blood vessels) twined around it. These are branches of the **efferent arteriole** (see **glomerulus**) and they re-unite to form larger blood vessels carrying blood from the **kidney**.

Collecting duct or collecting tubule
A tube which carries **urine** from several **uriniferous tubules** into the **pelvis** of a **kidney**.

Key to kidney diagram above

1. Glomerular filtration. *As blood squeezes through the* **glomerulus**, *most of its water, minerals, vitamins, glucose,* **amino acids*** *and* **urea** *are forced into the* **Bowman's capsule**, *forming* **glomerular filtrate**.

2. Glomerular filtrate *moves into* **proximal convoluted tubule**.

3. Tubular reabsorption. *As* **glomerular filtrate** *runs through the* **uriniferous tubule**, *most vitamins, glucose and* **amino acids*** *are taken back into the blood in the twining* **capillaries***.

4. *Some minerals are also taken back. The* **hormone* aldosterone*** *controls reabsorption of more if needed.*

5. *Some water is also taken back. The* **hormone* ADH*** *controls reabsorption of more if needed.*

6. Tubular secretion. *Some substances, e.g. ammonia and some drugs, pass from the blood into the* **uriniferous tubule**.

7. *Resulting* **urine** *passes into* **collecting duct**.

Distal *means "away from the point of origin or attachment".*

Proximal *means "near the point of origin or attachment".*

---

* **ADH, Aldosterone**, 108; **Amino acids**, 102 (**Proteins**); **Arteriole**, 60 (**Arteries**); **Capillaries**, 60; **Hormones**, 108; **Venule**, 60 (**Veins**).

THE CENTRAL NERVOUS SYSTEM

The **central nervous system (CNS)** is the body's control centre. It co-ordinates all its actions, both mechanical and chemical (working with **hormones***) and is made up of the **brain** and **spinal cord**. The millions of nerves in the body carry "messages" (nervous impulses) to and from these central areas (see pages 78-81).

Brain

Brain

The organ which controls most of the body's activities. It is the only organ able to produce "intelligent" action – action based on past experience (stored information), present events and future plans. It is made up of millions of **neurons*** (nerve cells), arranged into **sensory**, **association** and **motor areas**. The sensory areas receive information (nervous impulses) from all body parts and the association areas analyse the impulses and make decisions. The motor areas send impulses (orders) to muscles or glands. The impulses are carried by the fibres of 43 pairs of nerves – 12 pairs of **cranial nerves** serving the head, and 31 pairs of **spinal nerves** (see **spinal cord**).

Spinal cord (inside vertebral column*)

Spinal cord

A long string of nervous tissue running down from the **brain** inside the **vertebral column***. Nervous impulses from all parts of the body pass through it. Some are carried into or away from the brain, some are dealt with in the cord (see **involuntary actions**, page 81). 31 pairs of **spinal nerves** branch out from the cord through the gaps between the **vertebrae***. Each spinal nerve is made up of two groups of fibres: a **dorsal** or **sensory root**, made up of the fibres of **sensory neurons*** (bringing impulses in), and a **ventral** or **motor root**, made up of the fibres of **motor neurons*** (taking impulses out).

Vertebra*
Spinal cord
Spinal nerve

Neuroglia or glia

Stiffened cells which support and protect the nerve cells (**neurons***) of the central nervous system. Some produce a white, fatty substance called **myelin** (see also **Schwann cells**, page 76). This coats the long fibres found in the connective areas of the **brain** and the outer layer of the **spinal cord**, and leads to these areas being known as **white matter**. **Grey matter**, by contrast, consists mainly of **cell bodies*** and their short fibres, and its neuroglia do not produce myelin.

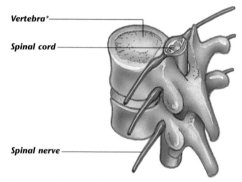

Spinal cord
Cerebrospinal fluid
Grey matter
Spinal nerve
Ventral root
Dorsal root
White matter

* **Cell body**, 76; **Hormones**, 108; **Motor neurons**, 77; **Neurons**, 76; **Sensory neurons**, 77; **Vertebrae**, **Vertebral column**, 51.

The parts of the brain

Cerebrum

The largest, most highly developed area, with many deep folds. It is composed of two **cerebral hemispheres**, joined by the **corpus callosum** (a band of **nerve fibres***), and its outer layer is called the **cerebral cortex**. This contains all the most important **sensory**, **association** and **motor areas** (see **brain**). It controls most physical activities and is the centre for mental activities such as decision-making, speech, learning, memory and imagination.

Cerebellum

The area which co-ordinates muscle movement and balance, two things under the overall control of the **cerebrum**.

Midbrain or mesencephalon

An area joining the **diencephalon** to the **pons**. It carries impulses in towards the **thalamus**, and out from the **cerebrum** towards the **spinal cord**.

Pons or pons Varolii

A junction of **nerve fibres*** which forms a link between the parts of the **brain** and the **spinal cord** (via the **medulla**).

Thalamus

The area which carries out the first, basic sorting of incoming impulses and directs them to different parts of the **cerebrum**. It also directs some outgoing impulses.

Hypothalamus

The master controller of most inner body functions. It controls the **autonomic nervous system*** (the nerve cells causing unconscious action, e.g. food movement through the intestines) and the action of the **pituitary gland***. Its activities are vital to **homeostasis*** – the maintenance of stable internal conditions.

Diencephalon

A collective term for the **thalamus** and **hypothalamus**.

Medulla or medulla oblongata

The area which controls the "fine tuning" of many unconscious actions (under the overall control of the **hypothalamus**). Different parts of it control different actions, e.g. the **respiratory centre** controls breathing.

Brain stem

A collective term for the **midbrain**, **pons** and **medulla**.

Brain (cut away)

Cerebrum

Thalamus

Cerebellum

Skull

Ventricles (spaces). Filled with CSF.

Hypothalamus

Pituitary gland*

Midbrain

Pons

Medulla

Corpus callosum

Protective membranes (**meninges**, sing. **meninx**). Called (working inwards) the **dura mater**, **arachnoid** and **pia mater**.

Cerebrospinal fluid (**CSF**) cushions the **brain** and **spinal cord** and brings dissolved food.

Spinal cord

In general terms, **cerebral** means "of the **brain**".

Cephalic means "of the head".

Areas of cerebrum

Sensory areas. Receive incoming impulses.
1. General sensory area. Receives impulses from muscles, skin and inner organs.
2. **Primary gustatory area**. Impulses from tongue.
3. **Primary auditory area**. Impulses from ears.
4. **Primary visual area**. Impulses from eyes.
5. **Primary olfactory area**. Impulses from nose.

Motor areas. Each tiny part sends out impulses to a specific muscle.

Association areas. Interpret impulses and make decisions. Some specific ones are:
6. **Visual association area**. Produces sight.
7. **Auditory association area**. Produces hearing.

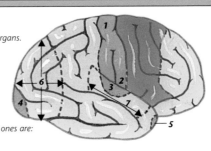

***Autonomic nervous system**, 80; **Homeostasis**, 107; **Nerve fibres**, 76; **Pituitary gland**, 69.

75

THE UNITS OF THE NERVOUS SYSTEM

The individual units of both the brain and spinal cord (**central nervous system***) and the nerves of the rest of the body (**peripheral nervous system**) are the nerve cells, or **neurons**. They are unique in being able to transmit electrical "messages" (the vital nervous impulses) around the body. Each neuron consists of a **cell body**, an **axon** and one or more **dendrites**, and there are three types of neuron – **sensory**, **association** and **motor neurons**.

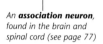

An **association neuron**, found in the brain and spinal cord (see page 77)

The parts of a neuron

Cell body or perikaryon
The part of a neuron containing its **nucleus*** and most of its **cytoplasm***. The cell bodies of all **association**, some **sensory** and some **motor neurons** lie in the brain and spinal cord. Those of the other sensory neurons are found in masses called **ganglia*** or as part of highly specialized **receptors*** in the nose and eyes. Those of the other motor neurons lie in **autonomic ganglia***.

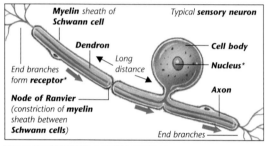

Myelin sheath of Schwann cell — Typical **sensory neuron**
Dendron — Cell body
Long distance — Nucleus*
End branches form **receptor***
Axon
Node of Ranvier (constriction of **myelin** sheath between Schwann cells)
End branches

Nerve fibres
The fibres (**axon** and **dendrites**) of a neuron. They are extensions of the **cytoplasm*** of the **cell body** and carry the vital nervous impulses. Most of the long nerve fibres which run out round the body (belonging to **sensory** or **motor neurons**) are accompanied by **neuroglial*** cells. These are called **Schwann cells** and they produce a sheath of **myelin*** around each fibre.

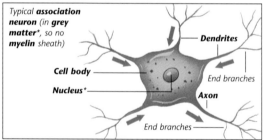

Typical **association neuron** (in **grey matter***, so no **myelin** sheath)
Dendrites
Cell body
End branches
Nucleus*
Axon
End branches

Dendrites
The **nerve fibres** carrying impulses towards a cell body. Most neurons have several short dendrites, but one type of **sensory neuron** has just one, elongated dendrite, often called a **dendron**. The endings of these dendrons form **receptors*** all over the body, and the dendrons themselves run inwards to the cell bodies (which are found in **ganglia*** just outside the spinal cord).

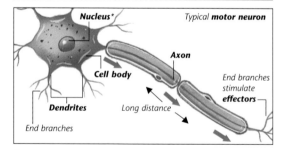

Nucleus* — Typical **motor neuron**
Axon
Cell body
End branches stimulate effectors
Dendrites
Long distance
End branches

Axon
The single long **nerve fibre** which carries impulses away from a **cell body**. The axons of all **association** and **sensory neurons** and some **motor neurons** lie in the brain and spinal cord. Those of the other motor neurons run out of the spinal cord to **autonomic ganglia***, or further to **effectors** (see **motor neurons**).

* **Autonomic ganglia**, 81; **Central nervous system**, 74; **Cytoplasm**, 10; **Ganglia**, 78; **Grey matter**, **Myelin**, 74 (**Neuroglia**); **Nucleus**, 10; **Receptors**, 79.

Types of neuron

Sensory neurons or afferent neurons

The neurons which carry "information" (nervous impulses) about sensations. The single **dendrites** (**dendrons**) of some sensory neurons run throughout the body, and their endings fire off impulses when stimulated. For more about these endings (**receptors**) and the different sensory neurons, see pages 78-79.

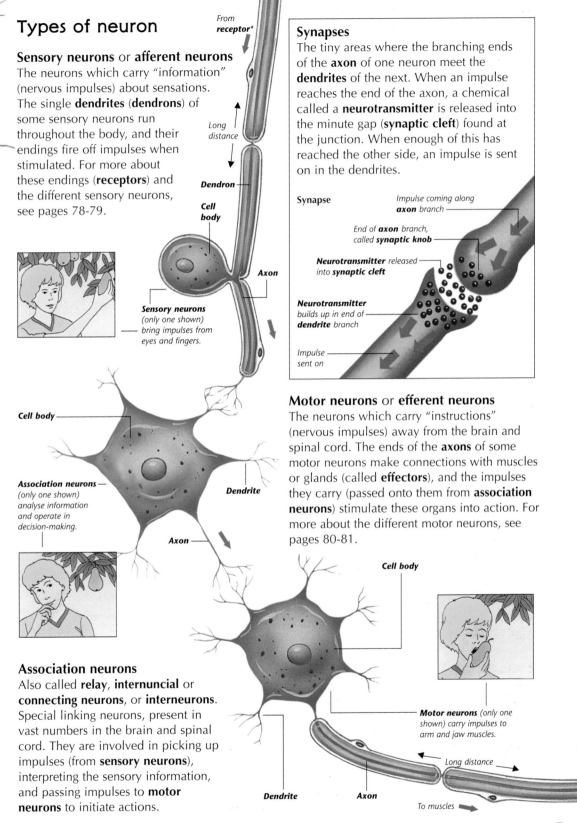

From receptor*

Long distance

Dendron

Cell body

Axon

Sensory neurons (only one shown) bring impulses from eyes and fingers.

Cell body

Association neurons (only one shown) analyse information and operate in decision-making.

Dendrite

Axon

Association neurons

Also called **relay**, **internuncial** or **connecting neurons**, or **interneurons**. Special linking neurons, present in vast numbers in the brain and spinal cord. They are involved in picking up impulses (from **sensory neurons**), interpreting the sensory information, and passing impulses to **motor neurons** to initiate actions.

Dendrite

Axon

Synapses

The tiny areas where the branching ends of the **axon** of one neuron meet the **dendrites** of the next. When an impulse reaches the end of the axon, a chemical called a **neurotransmitter** is released into the minute gap (**synaptic cleft**) found at the junction. When enough of this has reached the other side, an impulse is sent on in the dendrites.

Synapse

Impulse coming along axon branch

End of axon branch, called synaptic knob

Neurotransmitter released into synaptic cleft

Neurotransmitter builds up in end of dendrite branch

Impulse sent on

Motor neurons or efferent neurons

The neurons which carry "instructions" (nervous impulses) away from the brain and spinal cord. The ends of the **axons** of some motor neurons make connections with muscles or glands (called **effectors**), and the impulses they carry (passed onto them from **association neurons**) stimulate these organs into action. For more about the different motor neurons, see pages 80-81.

Cell body

Motor neurons (only one shown) carry impulses to arm and jaw muscles.

Long distance

To muscles

NERVES AND NERVOUS PATHWAYS

The **sensitivity** (**irritability**) of the body (its ability to respond to stimuli) relies on the transportation of "messages" (nervous impulses) by the fibres of nerve cells (**neurons***). The fibres which bring impulses into the brain and spinal cord are part of the **afferent system**. Those which carry impulses from the brain and cord are part of the **efferent system** (see pages 80-81). The fibres outside the brain and cord make up the **nerves** of the body, known collectively as the **peripheral nervous system** (**PNS**).

Nerves

Bundles of nerve fibres, blood vessels and **connective tissue***. Each nerve consists of several bundles (**fascicles**) of fibres and each fibre is part of a nerve cell (**neuron***). **Sensory nerves** have just the fibres (**dendrons***) of **sensory** (**afferent**) **neurons***, **motor nerves** have just the fibres (**axons***) of **motor** (**efferent**) **neurons***, and **mixed nerves** have both types of fibre.

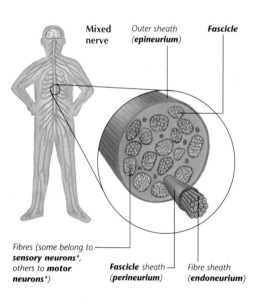

Mixed nerve | Outer sheath (**epineurium**) | **Fascicle**

Fibres (some belong to **sensory neurons***, others to **motor neurons***)

Fascicle sheath (**perineurium**) | Fibre sheath (**endoneurium**)

The afferent system

The **afferent system** is the system of nerve cells (**neurons***) whose fibres carry sensory information (nervous impulses) towards the spinal cord, up inside it and into the brain. The nerve cells involved are all the **sensory** (**afferent**) **neurons*** of the body. The impulses originate in **receptors** and are interpreted by the brain as sensations.

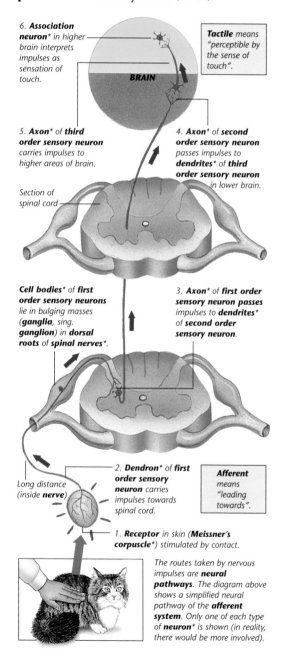

6. **Association neuron*** in higher brain interprets impulses as sensation of touch.

Tactile means "perceptible by the sense of touch".

BRAIN

5. **Axon*** of **third order sensory neuron** carries impulses to higher areas of brain.

4. **Axon*** of **second order sensory neuron** passes impulses to **dendrites*** of **third order sensory neuron** in lower brain.

Section of spinal cord

Cell bodies* of **first order sensory neurons** lie in bulging masses (**ganglia**, sing. **ganglion**) in **dorsal roots** of **spinal nerves***.

3. **Axon*** of **first order sensory neuron** passes impulses to **dendrites*** of **second order sensory neuron**.

2. **Dendron*** of **first order sensory neuron** carries impulses towards spinal cord.

Long distance (inside **nerve**)

Afferent means "leading towards".

1. **Receptor** in skin (**Meissner's corpuscle***) stimulated by contact.

The routes taken by nervous impulses are **neural pathways**. The diagram above shows a simplified neural pathway of the **afferent system**. Only one of each type of **neuron*** is shown (in reality, there would be more involved).

* **Association neurons**, 77; **Axon**, **Cell body**, 76; **Connective tissue**, 52; **Dendron**, 76 (**Dendrites**); **Meissner's corpuscles**, 83; **Motor neurons**, 77; **Neurons**, 76; **Sensory neurons**, 77; **Spinal nerves**, 74 (**Spinal cord**).

Receptors

The parts of the **afferent system** which fire off nervous impulses when they are stimulated. Most are either the single branched ending of the long **dendron*** of a **first order sensory neuron** (see picture) or a group of such endings. They are all embedded in body tissue, and many have some kind of structure formed around them (e.g. a **taste bud** – see **tongue**). They are found all over the body, both near the surface (in the skin, **sense organs**, **skeletal muscles***, etc.) and deeper inside (connected to inner organs, blood vessel walls, etc.).

Sense organs

The highly specialized sensory organs of the body, each with many **receptors**. They are the **nose**, **tongue**, eyes and ears. For more about eyes and ears, see pages 84-87.

Divisions of the afferent system

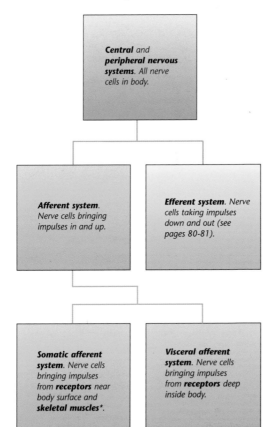

Central and **peripheral nervous systems**. All nerve cells in body.

Afferent system. Nerve cells bringing impulses in and up.

Efferent system. Nerve cells taking impulses down and out (see pages 80-81).

Somatic afferent system. Nerve cells bringing impulses from **receptors** near body surface and **skeletal muscles***.

Visceral afferent system. Nerve cells bringing impulses from **receptors** deep inside body.

Nose

The organ of smell. Each of its two nostrils opens out into a **nasal cavity** which is lined with **mucous membrane*** and has many **olfactory hairs** extending from its roof. The hairs are the **dendrites*** of special **sensory neurons*** called **olfactory cells**. These are the **receptors** whose impulses are interpreted by the brain as sensations of smell (**olfactory sensations**).

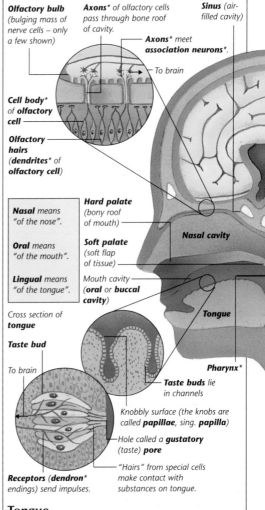

Olfactory bulb (bulging mass of nerve cells – only a few shown)

Axons* of olfactory cells pass through bone roof of cavity.

Sinus (air-filled cavity)

Axons* meet **association neurons***.

To brain

Cell body* of **olfactory cell**

Olfactory hairs (**dendrites*** of **olfactory cell**)

Nasal means "of the nose".

Oral means "of the mouth".

Lingual means "of the tongue".

Hard palate (bony roof of mouth)

Soft palate (soft flap of tissue)

Mouth cavity (**oral** or **buccal cavity**)

Nasal cavity

Tongue

Pharynx*

Cross section of **tongue**

Taste bud

To brain

Taste buds lie in channels

Knobbly surface (the knobs are called **papillae**, sing. **papilla**)

Hole called a **gustatory** (taste) **pore**

"Hairs" from special cells make contact with substances on tongue.

Receptors (**dendron*** endings) send impulses.

Tongue

The main organ of taste. It is a muscular organ which bears many **taste buds**. These tiny bodies contain the **receptors** whose impulses are interpreted by the brain as taste sensations (**gustatory sensations**).

* **Association neurons**, 77; **Axon, Cell body**, 76; **Dendron**, 76 (**Dendrites**); **Mucous membrane**, 67; **Pharynx**, 66; **Sensory neurons**, 77; **Skeletal muscles**, 54.

The efferent system

The **efferent system** is the second system of nerve cells (**neurons***) in the body (see also **afferent system**, pages 78-79). The fibres of its nerve cells carry nervous impulses away from the brain, down through the spinal cord and out around the body. The nerve cells involved are all the **motor (efferent) neurons*** of the body. The impulses they carry stimulate action in the surface muscles (**skeletal muscles***) or in the glands and internal muscles (in the walls of inner organs and blood vessels). All these organs are known collectively as **effectors**.

Divisions of the efferent system

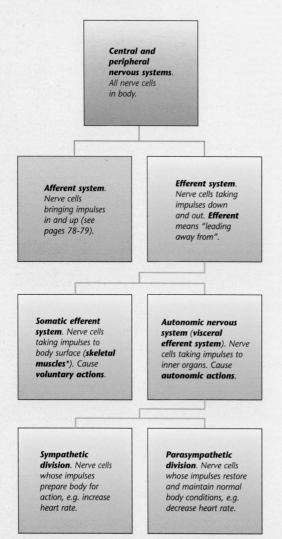

Central and peripheral nervous systems. All nerve cells in body.

Afferent system. Nerve cells bringing impulses in and up (see pages 78-79).

Efferent system. Nerve cells taking impulses down and out. **Efferent** means "leading away from".

Somatic efferent system. Nerve cells taking impulses to body surface (**skeletal muscles***). Cause **voluntary actions**.

Autonomic nervous system (visceral efferent system). Nerve cells taking impulses to inner organs. Cause **autonomic actions**.

Sympathetic division. Nerve cells whose impulses prepare body for action, e.g. increase heart rate.

Parasympathetic division. Nerve cells whose impulses restore and maintain normal body conditions, e.g. decrease heart rate.

The different actions

Voluntary actions

Actions which result from conscious activity by the brain, i.e. ones it consciously decides upon, e.g. lifting a cup. We are always aware of these actions, which involve **skeletal muscles*** only. The impulses which cause them originate in higher areas of the brain (especially the **cerebrum***) and are carried by nerve cells of the **somatic efferent system**.

Lifting a cup is a **voluntary action**.

Simplified neural pathway* of voluntary action (somatic efferent system)

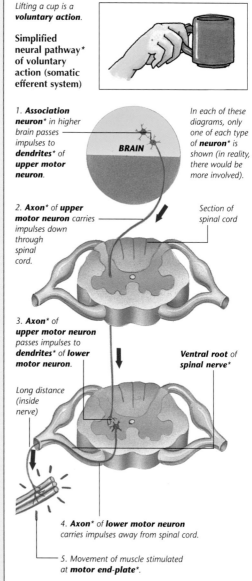

1. **Association neuron*** in higher brain passes impulses to **dendrites*** of **upper motor neuron**.

In each of these diagrams, only one of each type of **neuron*** is shown (in reality, there would be more involved).

2. **Axon*** of **upper motor neuron** carries impulses down through spinal cord.

Section of spinal cord

BRAIN

3. **Axon*** of **upper motor neuron** passes impulses to **dendrites*** of **lower motor neuron**.

Long distance (inside nerve)

Ventral root of spinal nerve*

4. **Axon*** of **lower motor neuron** carries impulses away from spinal cord.

5. Movement of muscle stimulated at **motor end-plate***.

* **Association neurons**, 77; **Axon**, 76; **Cerebrum**, 75; **Dendrites**, 76; **Motor end-plate**, 55; **Motor neurons**, 77; **Neural pathways**, 78; **Neurons**, 76; **Skeletal muscles**, 54; **Spinal nerves**, 74 (**Spinal cord**).

Involuntary actions

Automatic actions (ones the brain does not consciously decide upon). There are two types. Firstly, there are the constant actions of inner organs, e.g. the beating of the heart, of which we are not normally aware. The impulses which cause them originate in the lower brain (especially the **hypothalamus***) and are carried by nerve cells of the **autonomic nervous system**. They are called **autonomic actions**. The other involuntary actions are **reflex actions**.

The heartbeat is an ***autonomic action***.

Simplified neural pathway* of autonomic action (sympathetic division of autonomic nervous system)

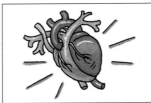

1. **Association neuron*** in lower brain passes impulses to **dendrites*** of **upper motor neuron**.

2. **Axon*** of **upper motor neuron** carries impulses down through spinal cord.

3. **Axon*** of **upper motor neuron** passes impulses to **dendrites*** of **preganglionic motor neuron**.

Section of spinal cord

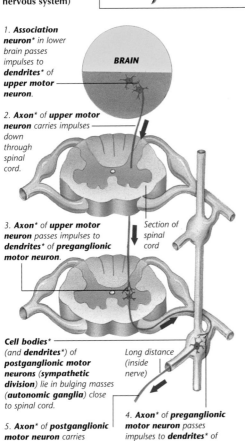

Cell bodies* (and **dendrites***) of **postganglionic motor neurons** (sympathetic division) lie in bulging masses (**autonomic ganglia**) close to spinal cord.

Long distance (inside nerve)

5. **Axon*** of **postganglionic motor neuron** carries impulses to organ.

4. **Axon*** of **preganglionic motor neuron** passes impulses to **dendrites*** of **postganglionic motor neuron**.

Reflex actions

Involuntary actions of which we are aware. The term is most often used to refer to sudden actions of **skeletal muscles***, e.g. snatching the hand away from something hot. The impulses which cause such an action are carried by nerve cells of the **somatic efferent system** and the entire **neural pathway*** is a "short-circuited" one, called a **reflex arc**. In the case of **cranial reflexes** (those of the head, e.g. sneezing), this pathway involves a small part of the brain; with **spinal reflexes** (those of the rest of the body), the brain is not actively involved, only the spinal cord.

Pulling your hand away from an intense source of heat is a ***reflex action***.

Simplified reflex arc (spinal reflex)

1. **Pain receptor*** stimulated

2. **Dendron*** of **first order sensory neuron** (see page 78) carries impulses to spinal cord.

Axon* of **second order sensory neuron** (see page 78) carries impulses to brain to "tell" it what has happened.

Long distance (inside nerve)

3. **Axon*** of **first order sensory neuron** passes impulses to **dendrites*** of **association neuron***.

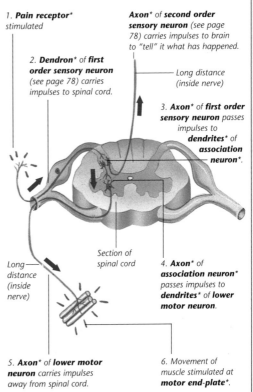

Long distance (inside nerve)

Section of spinal cord

4. **Axon*** of **association neuron*** passes impulses to **dendrites*** of **lower motor neuron**.

5. **Axon*** of **lower motor neuron** carries impulses away from spinal cord.

6. Movement of muscle stimulated at **motor end-plate***.

* **Association neurons**, 77; **Axon**, **Cell body**, 76; **Dendron**, 76 (**Dendrites**); **Hypothalamus**, 75; **Motor end-plate**, 55; **Neural pathways**, 78; **Pain receptors**, 83; **Skeletal muscles**, 54.

81

THE SKIN

The **skin** or **cutis** is the outer body covering, made up of several tissue layers. It registers external stimulation, protects against damage or infection, prevents drying out, helps regulate body temperature, excretes waste (**sweat**), stores fat and makes **vitamin D***. It contains many tiny structures, each type with a different function. The entire skin (tissue layers and structures), is called the **integumentary system**.

The different layers

Epidermis

The thin outer layer of the skin which forms its **epithelium** (a term for any sheet of cells which forms a surface covering or a cavity lining). It is made up of several layers (**strata**, sing. **stratum**), shown in the picture, right.

*1. **Stratum corneum** (**horny** or **cornified layer**). Flat, dead cells filled with **keratin** (a fibrous, waterproofing protein). The cells are continually worn away or shed.*

*2. **Stratum granulosum** (**granular layer**). Flat, granulated cells. They slowly die away (there are no blood vessels in the epidermis to provide food and oxygen) and are pushed up to become part of the **stratum corneum**.*

*3. **Stratum germinativum**. Made up of two layers. The upper one (**stratum spinosum**) consists of new living cells. These push upwards (and become part of the **stratum granulosum**) as more cells are made below them by the constantly-dividing cells of the lower layer (**stratum basale** or **Malphigian layer**).*

Dermis

The thick layer of **connective tissue*** under the **epidermis**, containing most of the embedded structures (see introduction). It also contains many **capillaries*** (tiny blood vessels) which supply food and oxygen.

Epidermal layers

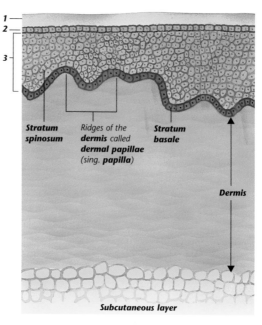

Stratum spinosum | Ridges of the **dermis** called **dermal papillae** (sing. **papilla**) | **Stratum basale**

Dermis

Subcutaneous layer

Subcutaneous layer or superficial fascia

The layer of fatty tissue (**adipose tissue**) below the **dermis** (it is a fat store). Elastic fibres run through it to connect the dermis to the organs below, e.g. muscles. It forms an insulating layer. See also picture, right.

Melanin

A brown **pigment*** which shields against ultraviolet light by absorbing the light energy. It is found in all the layers of the **epidermis** of people from tropical areas, giving them dark skin. Fair-skinned people only have melanin in their lower epidermal layers, but produce more when in direct sunlight, causing a suntan.

*People with **melanin** only in lower layers of the **epidermis** have fair skin.*

*The **pigment*** **carotene**, together with **melanin**, causes yellow skin.*

*Dark skin is caused by large amounts of **melanin** in all epidermal layers.*

---

* **Capillaries**, 60; **Connective tissue**, 52; **Pigments**, 27; **Vitamin D**, 111.

Structures in the skin

1. Meissner's corpuscles
Bodies formed around nerve fibre endings. There are especially large numbers on the fingertips and palms. They are touch **receptors***, i.e. they send impulses to the brain when the skin makes contact with an object.

2. Sebaceous glands
Exocrine glands* which open into **hair follicles**. They produce an oil called **sebum** which waterproofs the hairs and **epidermis** and keeps them supple.

3. Hair erector muscles
Special muscles, each attached to a **hair follicle**. When they contract (in the cold), the hairs straighten. This traps more air and improves insulation (especially in animals with lots of hair, feathers or fur). It also causes "goose-pimples".

4. Hair follicles
Long, narrow tubes, each containing a hair. The hair grows as new cells are added to its base from the cells lining the follicle. Its older cells die as **keratin** forms inside them (see **stratum corneum**).

5. Pain receptors
Nerve fibre endings in the tissue of most inner organs and in the skin (in the **epidermis** and the top of the **dermis**). They are the **receptors*** which send impulses when any stimulation (e.g. pressure, heat, touch) becomes excessive. This is what causes a sensation of pain.

6. Hair plexuses or root hair plexuses
Special groups of nerve fibre endings. Each forms a network around a **hair follicle** and is a **receptor***, i.e. it sends nervous impulses to the brain, in this case when the hair moves.

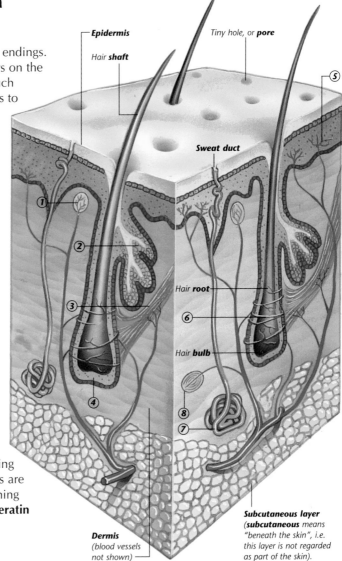

Epidermis

Hair **shaft**

Tiny hole, or **pore**

Sweat duct

Hair **root**

Hair **bulb**

Dermis
(blood vessels not shown)

Subcutaneous layer
(**subcutaneous** means "beneath the skin", i.e. this layer is not regarded as part of the skin).

7. Sweat glands or sudoriferous glands
Coiled **exocrine glands*** which excrete **sweat**. Each has a narrow tube (**sweat duct**) going to the surface. Sweat consists of water, salts and **urea***, which enter the gland from the cells and **capillaries*** (blood vessels).

8. Pacinian corpuscles
Special bodies formed around single nerve fibre endings, lying in the lower skin layers and the walls of inner organs. They are pressure **receptors***, i.e. they send impulses to the brain when the tissue is receiving deep pressure rather than light touch.

* **Capillaries**, 60; **Exocrine glands**, 68; **Receptors**, 79; **Urea**, 72.

THE EYES

The **eyes** are the organs of **sight**, sending nervous impulses to the brain when stimulated by light rays from external objects. The brain interprets the impulses to produce images. Each eye consists of a hollow, spherical capsule (**eyeball**), made up of several layers and structures. It is set into a socket in the skull (an **orbit**), and is protected by eyelids and eyelashes.

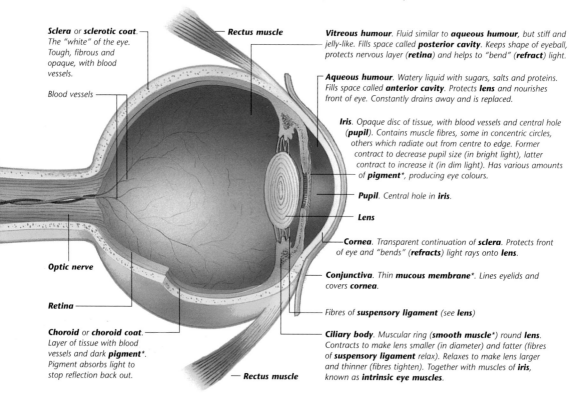

Sclera or **sclerotic coat**. The "white" of the eye. Tough, fibrous and opaque, with blood vessels.

Blood vessels

Rectus muscle

Optic nerve

Retina

Choroid or **choroid coat**. Layer of tissue with blood vessels and dark **pigment***. Pigment absorbs light to stop reflection back out.

Rectus muscle

Vitreous humour. Fluid similar to **aqueous humour**, but stiff and jelly-like. Fills space called **posterior cavity**. Keeps shape of eyeball, protects nervous layer (**retina**) and helps to "bend" (**refract**) light.

Aqueous humour. Watery liquid with sugars, salts and proteins. Fills space called **anterior cavity**. Protects **lens** and nourishes front of eye. Constantly drains away and is replaced.

Iris. Opaque disc of tissue, with blood vessels and central hole (**pupil**). Contains muscle fibres, some in concentric circles, others which radiate out from centre to edge. Former contract to decrease pupil size (in bright light), latter contract to increase it (in dim light). Has various amounts of **pigment***, producing eye colours.

Pupil. Central hole in **iris**.

Lens

Cornea. Transparent continuation of **sclera**. Protects front of eye and "bends" (**refracts**) light rays onto **lens**.

Conjunctiva. Thin **mucous membrane***. Lines eyelids and covers **cornea**.

Fibres of **suspensory ligament** (see **lens**)

Ciliary body. Muscular ring (**smooth muscle***) round **lens**. Contracts to make lens smaller (in diameter) and fatter (fibres of **suspensory ligament** relax). Relaxes to make lens larger and thinner (fibres tighten). Together with muscles of **iris**, known as **intrinsic eye muscles**.

Lens

The transparent body in an eye, whose role, like that of any lens, is to focus the light rays passing through it, i.e. to "bend" (**refract**) them so that they come to a point, in this case on the **retina**. A lens consists of many thin tissue layers and is held in place by the fibres of a **ligament*** called the **suspensory ligament**. These join it to the **ciliary body**, which can alter the lens shape so that light rays are always focused on the retina, whatever the distance of the object being looked at. This is known as **accommodation**. The rays form an upside-down image on the retina, but this is corrected by the brain, so that we "see" things the right way up.

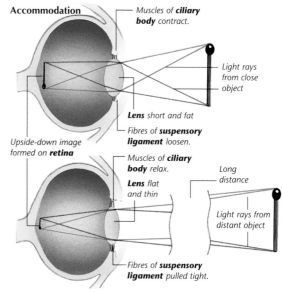

Accommodation

Muscles of **ciliary body** contract.

Light rays from close object

Lens short and fat

Fibres of **suspensory ligament** loosen.

Upside-down image formed on **retina**

Muscles of **ciliary body** relax.

Lens flat and thin

Long distance

Light rays from distant object

Fibres of **suspensory ligament** pulled tight.

***Ligaments**, 52; **Mucous membrane**, 67; **Pigments**, 27; **Smooth muscle**, 55.

The inner nervous layer

Retina

The innermost layer of tissue at the back of the eyeball, made up of a layer of **pigment*** and a nervous layer consisting of millions of sensory nerve cells (**sensory neurons***) and their fibres. These lie in chains and carry nervous impulses to the brain. The first cells in the chains are **receptors***, i.e. their end fibres (**dendrons***) fire off the impulses when they are stimulated (by light rays). These fibres are called **rods** and **cones** because of their shapes. The receptors are **photoreceptors** (i.e. stimulated by light).

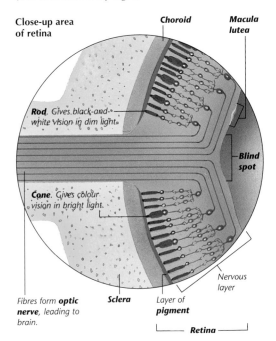

Close-up area of retina

Choroid

Macula lutea

Rod. *Gives black-and-white vision in dim light*.

Blind spot

Cone. *Gives colour vision in bright light.*

Nervous layer

*Fibres form **optic nerve**, leading to brain.*

Sclera

Layer of **pigment**

Nervous layer

└── **Retina** ──┘

Macula lutea or yellowspot

An area of yellowish tissue in the centre of the **retina**. It has a small central dip, called the **fovea** or **fovea centralis**. This has the highest concentration of **cones** (see **retina**) and is the area of acutest vision. If you look directly at a specific object, its light rays are focused on the fovea.

Blind spot or optic disc

The point in the **retina** where the **optic nerve** leaves the eye. It has no **receptors** (see **retina**) and so cannot send any impulses.

Structures around the eyeballs

Extrinsic eye muscles

The three pairs of muscles joining the eyeball to the eye socket (**orbit**). They contract to make the eyeball swivel around.

Extrinsic eye muscles

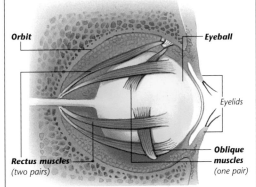

Orbit

Eyeball

Eyelids

Rectus muscles *(two pairs)*

Oblique muscles *(one pair)*

Lachrymal glands or tear glands

Two **exocrine glands***, one at the top of each eye socket (**orbit**). They secrete a watery fluid onto the lining of the upper eyelids via tubes called **lachrymal ducts**. The fluid contains salts and an anti-bacterial **enzyme***, and it washes over the surface of the eyes, keeping them moist and clean. It drains away via four **lachrymal canals**, two at the inside corner of each eye, which join to form a **nasolachrymal duct**. This empties into a **nasal cavity***.

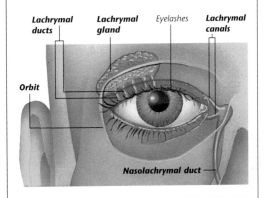

Lachrymal ducts

Lachrymal gland

Eyelashes

Lachrymal canals

Orbit

Nasolachrymal duct

Optic *means "of vision or the eye".*

Visual *means "perceptible by the sense of sight".*

* **Dendron**, 76 (**Dendrites**); **Enzymes**, 105; **Exocrine glands**, 68;
Nasal cavity, 79 (**Nose**); **Pigments**, 27; **Receptors**, 79; **Sensory neurons**, 77.

THE EARS

The two **ears** are the organs of hearing and balance. Each one is divided into three areas – the **outer ear**, the **middle ear** and the **inner ear**.

Outer ear or external ear
An outer "shell" of skin and **cartilage** * (**pinna** or **auricle**), together with a short tube (**ear canal** or **external auditory canal**). The tube lining contains special **sebaceous glands*** (**ceruminous glands**) which secrete **cerumen** (ear wax).

Middle ear or tympanic cavity
An air-filled cavity which contains a chain of three tiny bones (**ear ossicles or auditory ossicles**) called the **malleus** (or **hammer**), **incus** (or **anvil**) and **stapes** (or **stirrup**).

Inner ear or internal ear
A connected series of cavities in the skull, with tubes and sacs inside them. The cavities (**cochlea, vestibule** and **semicircular canals**) are called the **bony labyrinth** and are filled with one fluid (**perilymph**). The tubes and sacs are filled with another fluid (**endolymph**) and are called the **membranous labyrinth**. They are the **cochlear duct, saccule, utricle** and **semicircular ducts**.

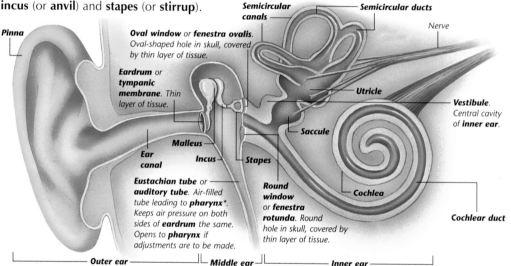

Semicircular canals

Semicircular ducts

Nerve

Pinna

Oval window or **fenestra ovalis**. Oval-shaped hole in skull, covered by thin layer of tissue.

Eardrum or **tympanic membrane**. Thin layer of tissue.

Utricle

Vestibule. Central cavity of **inner ear**.

Malleus

Saccule

Ear canal

Incus

Stapes

Eustachian tube or **auditory tube**. Air-filled tube leading to **pharynx***. Keeps air pressure on both sides of **eardrum** the same. Opens to **pharynx** if adjustments are to be made.

Round window or **fenestra rotunda**. Round hole in skull, covered by thin layer of tissue.

Cochlea

Cochlear duct

⌞ **Outer ear** ⌟⌞ **Middle ear** ⌟⌞ **Inner ear** ⌟

The inner ear and hearing

Cochlea
A spiralling tubular cavity, part of the **inner ear**. It contains **perilymph** (see **inner ear**) in two channels (continuous with each other), and also a third channel – the **cochlear duct**.

Cochlear duct
A spiralling tube within the **cochlea**, connected to the **saccule**. It contains **endolymph** (see **inner ear**) and a long body called the **organ of Corti**. This contains special hair cells whose hairs project into the endolymph and touch a shelf-like tissue layer (**tectorial membrane**). The bases of the cells are attached to nerve fibres (**dendron*** endings).

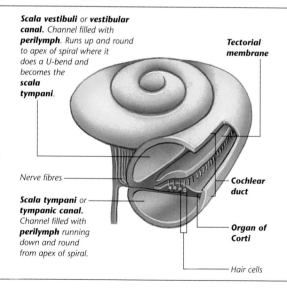

Scala vestibuli or **vestibular canal**. Channel filled with **perilymph**. Runs up and round to apex of spiral where it does a U-bend and becomes the **scala tympani**.

Tectorial membrane

Nerve fibres

Scala tympani or **tympanic canal**. Channel filled with **perilymph** running down and round from apex of spiral.

Cochlear duct

Organ of Corti

Hair cells

* **Cartilage**, 53; **Dendron**, 76 (**Dendrites**); **Pharynx**, 66; **Sebaceous glands**, 83.

The inner ear and balance

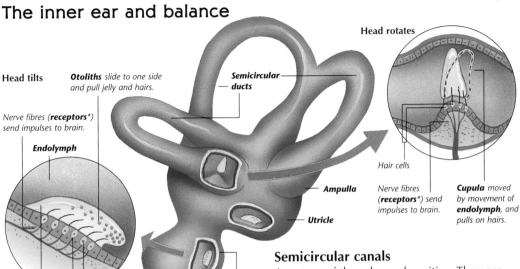

Head tilts

Otoliths *slide to one side and pull jelly and hairs.*

Nerve fibres (**receptors***) *send impulses to brain.*

Endolymph

Semicircular ducts

Head rotates

Hair cells

Nerve fibres (**receptors***) *send impulses to brain.*

Cupula *moved by movement of* **endolymph**, *and pulls on hairs.*

Ampulla

Utricle

Macula

Hair cells

Saccule

Saccule (sacculus) and utricle (utriculus)

Two sacs lying between the **semicircular ducts** and the **cochlear duct**. They contain **endolymph** (see **inner ear**) and have special hair cells in patches in their linings. These cells have nerve fibres (**dendron*** endings) attached to them and hairs embedded in a jelly-like mass called a **macula** (pl. **maculae**). This contains grains of calcium carbonate (**otoliths**). The maculae send the brain information about forward, backward, sideways or tilting motion of the head.

Semicircular canals

A system of three looped cavities. They are part of the **inner ear** and are positioned on the three different planes of movement, at right angles to each other.

Semicircular ducts

Three looped tubes inside the **semicircular canals**. Each contains **endolymph** (see **inner ear**) and a special sensory body, which lies across the basal swelling (**ampulla**, pl. **ampullae**) of the duct. The sensory bodies (**cupulae**, sing. **cupula**) work in a very similar way to **maculae** (see **saccule**) – each consists of a jelly-like mass (without **otoliths**) and hair cells. They send the brain information about rotation and tilting of the head.

a) *Sound waves (air vibrations) come in along* **ear canal** *and make* **eardrum** *vibrate.*

b) **Ear ossicles** *pick up vibrations and pass them to* **oval window** *(lever action magnifies vibrations about 20 times).*

c) *Vibrations of* **oval window** *cause waves in* **perilymph** *of* **vestibule**.

d) *Waves in* **perilymph** *of* **scala vestibuli** *cause waves in* **endolymph** *of* **cochlear duct**.

e) *Hairs move and cause nerve fibres (***receptors****) to send impulses to brain (which interprets them as sensation of hearing).*

f) *Waves gradually fade out.*

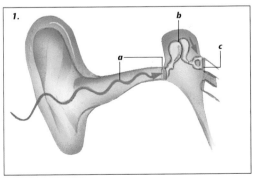

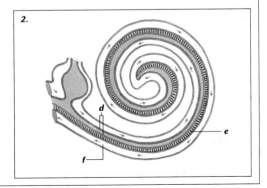

* **Dendron**, 76 (**Dendrites**);
  Receptors, 79.

THE REPRODUCTIVE SYSTEM

Reproduction is the process of producing new life. Humans reproduce by **sexual reproduction*** (described on pages 90-91) and the reproductive organs involved (making up the **reproductive system**) are called the **genital organs** or **genitalia**. They consist of the primary reproductive organs, or **gonads** (two **ovaries** in women, two **testes** in men) and a number of additional organs. In both women and men, cells in the gonads also act as **endocrine glands***, secreting many important **hormones***.

The male reproductive system

Testes (sing. **testis**) or **testicles**
The two male **gonads** (see introduction). They contain tube-like canals called **seminiferous tubules**, inside which the male **gametes*** (sex cells), called **sperm**, are made after **puberty*** (for more about how sperm are made, see pages 94-95). The testes lie in a sac (**scrotum**), which hangs below the abdomen (the temperature for sperm production must be slightly lower than body temperature). They also produce **hormones*** (**androgens** – see pages 108-109) after puberty.

Side view of male organs
(only one testis shown)

Sperm duct   **Bladder***

Urethra*

Penis

Sperm duct or **vas deferens** (pl. **vasa deferentia**). Continuation of **epididymis**, carrying sperm into **urethra*** during **ejaculation***.

Scrotum   **Anus***

Loose skin over **glans** called **foreskin** or **prepuce**

Testis

Glans. Tip of penis (most sensitive part) with many blood vessels.

Cross section of testis

Epididymis (pl. **epididymides**). Comma-shaped organ enclosing coiled tube where **sperm** are stored.

Interstitial cells (cells between tubules)

Seminiferous tubule

Ducts and glands
(view from behind – testes not shown)

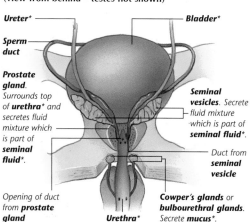

Ureter*   **Bladder***

Sperm duct

Prostate gland. Surrounds top of **urethra*** and secretes fluid mixture which is part of **seminal fluid***.

Seminal vesicles. Secrete fluid mixture which is part of **seminal fluid***.

Duct from seminal vesicle

Opening of duct from **prostate gland**   **Urethra***   **Cowper's glands** or **bulbourethral glands**. Secrete **mucus***.

Penis
The organ through which **sperm** (see **testes**) are ejected (via the **urethra***) during **copulation***. It is made of soft, sponge-like **erectile tissue**, which has many spaces (**blood sinuses**), blood vessels and nerve fibre endings (**receptors***). When a man is sexually excited, the sinuses and blood vessels fill with blood (the blood vessels expand). This makes the penis stiff and erect.

* **Anus**, 67 (**Large intestine**); **Bladder**, 72; **Ejaculation**, 91 (**Copulation**); **Endocrine glands**, 69; **Gametes**, 92; **Hormones**, 108; **Mucus**, 67 (**Mucous membrane**); **Puberty**, 90; **Receptors**, 79; **Seminal fluid**, 91 (**Copulation**); **Sexual reproduction**, 92; **Ureters, Urethra**, 72.

The female reproductive system

Ovaries

The two female **gonads** (see introduction). They are held in place in the lower abdomen (below the kidneys) by **ligaments***. These attach them to the walls of the pelvis. The female **gametes*** (sex cells), called **ova** (sing. **ovum**), are produced regularly in the ovaries (in **ovarian follicles**) after **puberty***. For more about how ova are made, see pages 94-95.

Vulva or **pudendum**

A collective term for the outer parts of the female reproductive system – the **labia** and the **clitoris** (see picture, bottom right). The **labia** (**majora** and **minora**) are two folds of skin (one inside the other) which surround the openings from the **vagina** and the **urethra***. The clitoris is the most sensitive part. Like the **penis**, it is made of **erectile tissue** and has many **receptors***.

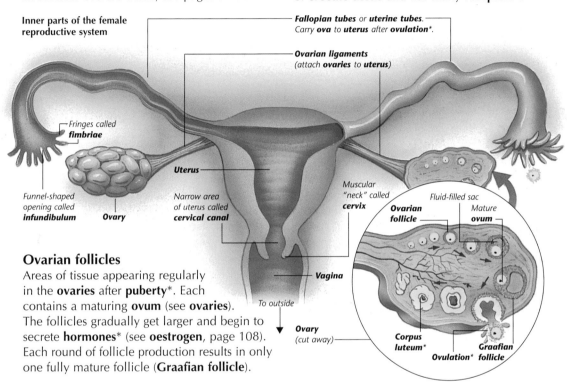

Inner parts of the female reproductive system

Fallopian tubes or uterine tubes. Carry ova to uterus after ovulation*.

Ovarian ligaments (attach ovaries to uterus)

Fringes called fimbriae

Funnel-shaped opening called infundibulum

Ovary

Uterus

Narrow area of uterus called cervical canal

Muscular "neck" called cervix

Ovarian follicle

Fluid-filled sac

Mature ovum

Vagina

To outside

Ovary (cut away)

Corpus luteum*

Ovulation* follicle

Graafian follicle

Ovarian follicles

Areas of tissue appearing regularly in the **ovaries** after **puberty***. Each contains a maturing **ovum** (see **ovaries**). The follicles gradually get larger and begin to secrete **hormones*** (see **oestrogen**, page 108). Each round of follicle production results in only one fully mature follicle (**Graafian follicle**).

Uterus or **womb**

The hollow organ, inside which a developing baby (**foetus***) is held, or from which the **ova** (see **ovaries**) are discharged (see **menstrual cycle**, page 90). It has a lining of **mucous membrane*** (the **endometrium**), covering a muscular wall with many blood vessels.

Vagina

The muscular canal leading from the **uterus** out of the body. It carries away the **ova** (see **ovaries**) and **endometrium** (see uterus) during **menstruation***, receives the **penis** during **copulation*** and serves as the birth canal. Its lining produces a lubricating fluid.

Outer parts of the female reproductive system

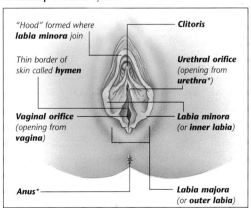

"Hood" formed where labia minora join

Thin border of skin called hymen

Vaginal orifice (opening from vagina)

Anus*

Clitoris

Urethral orifice (opening from urethra*)

Labia minora (or inner labia)

Labia majora (or outer labia)

* **Anus**, 67 (Large intestine); **Copulation**, 91; **Corpus luteum**, 90 (**Menstrual cycle**); **Foetus**, 91 (**Pregnancy**); **Gametes**, 92; **Hormones**, 108; **Ligaments**, 52; **Menstruation**, 90 (**Menstrual cycle**); **Mucous membrane**, 67; **Ovulation**, 90 (**Menstrual cycle**); **Puberty**, 90; **Receptors**, 79; **Urethra**, 72.

89

DEVELOPMENT AND REPRODUCTION

Humans reproduce by **sexual reproduction***. The main processes this involves are described on these two pages, as well as the initial developments which allow it to happen.

Puberty
The point when the reproductive organs mature, and a person becomes capable of reproducing – roughly between the ages of 11 and 15 in girls, and 13 and 15 in boys. It involves a number of significant changes, all stimulated by **hormones*** (see **oestrogen** and **androgens**, pages 108-109). All the new resulting features are called **secondary sex characters**, as distinct from the **primary sex characters** – the sex organs present from birth (see pages 88-89).

Menstrual cycle
A series of preparatory changes in the **uterus*** lining (**endometrium**), in case of **fertilization**. The lining gradually develops a new inner layer rich in blood vessels. If a fertilized **ovum** (female sex cell) does not appear, this new layer breaks down and leaves the body via the **vagina*** (**menstruation**). Each menstrual cycle lasts about 28 days and they occur continuously from **puberty** (usually between the ages of 11 and 15 – see left) to **menopause** (usually between 45 and 50), when ova production ceases. The events of the menstrual cycle run in conjunction with the **ovarian cycle** – the regular maturation of an ovum in an **ovarian follicle***, followed by **ovulation** (the release of the ovum into a **Fallopian tube***), and the breakdown of the **corpus luteum**. This body is formed from the burst **Graafian follicle*** (it does not break down if an ovum is fertilized). Both cycles are controlled by a group of **hormones*** (see pages 108-109).

Changes at puberty

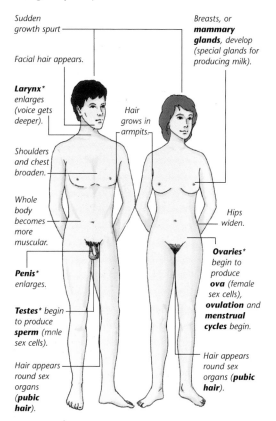

Sudden growth spurt

Facial hair appears.

Larynx* enlarges (voice gets deeper).

Shoulders and chest broaden.

Whole body becomes more muscular.

Penis* enlarges.

Testes* begin to produce **sperm** (male sex cells).

Hair appears round sex organs (**pubic hair**).

Breasts, or **mammary glands**, develop (special glands for producing milk).

Hair grows in armpits.

Hips widen.

Ovaries* begin to produce **ova** (female sex cells), **ovulation** and **menstrual cycles** begin.

Hair appears round sex organs (**pubic hair**).

Menstrual cycle / Ovarian cycle

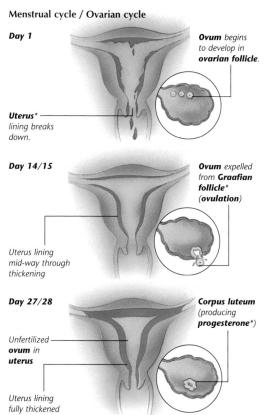

Day 1

Uterus* lining breaks down.

Ovum begins to develop in *ovarian follicle*.

Day 14/15

Uterus lining mid-way through thickening

Ovum expelled from *Graafian follicle** (*ovulation*)

Day 27/28

Unfertilized **ovum** in **uterus**

Uterus lining fully thickened

Corpus luteum (producing **progesterone***)

Copulation

Also called **coitus** or **sexual intercourse**. The insertion of the **penis*** into the **vagina***, followed by rhythmical movements of the pelvis in one or both sexes. Its culmination in the male is **ejaculation** – the ejection of **semen** from the **urethra*** (in the penis) into the vagina. Semen consists of **sperm** (male sex cells) and a fluid mixture (**seminal fluid**).

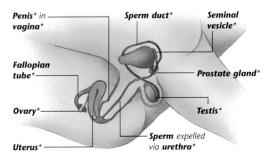

Penis* in vagina*

Sperm duct*

Seminal vesicle*

Fallopian tube*

Prostate gland*

Ovary*

Testis*

Sperm expelled via **urethra***

Uterus*

Fertilization

A process which occurs after **ejaculation** if the sperm (male sex cells) meet an **ovum** (female sex cell) in a **Fallopian tube***. One sperm penetrates the ovum's outer skin (**zona pellucida**). Its **nucleus*** fuses with that of the ovum, and the first cell of a new baby (**zygote***) is formed. The new cell travels towards the **uterus***, undergoing many cell divisions (**cleavage***) as it does so. The ball of cells formed from these divisions then becomes embedded in the uterus wall (**implantation**), after which it is called an **embryo***.

Zona pellucida

Sperm penetrates **ovum**. **Nucleus*** will fuse with ovum nucleus.

"Tail" is left behind.

Pregnancy

Pregnancy, or **gestation**, is the state of carrying young. The time between **fertilization** and giving birth (**parturition**) is the **gestation period** (about 9 months in humans) and the new developing individual in the **uterus*** is called a **foetus**, a term usually used instead of **embryo*** after about two months of pregnancy. A series of powerful muscular contractions called **labour** occur just before parturition.

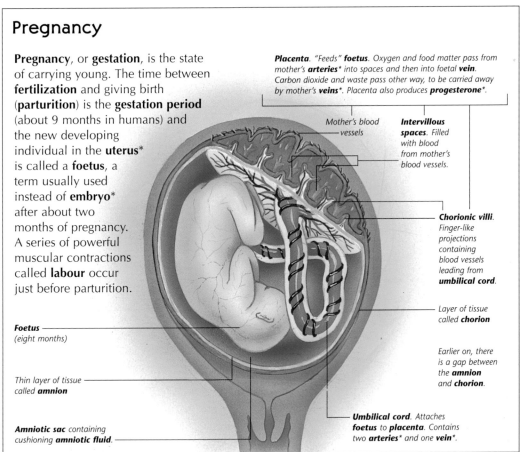

Placenta. "Feeds" **foetus**. Oxygen and food matter pass from mother's **arteries*** into spaces and then into foetal **vein**. Carbon dioxide and waste pass other way, to be carried away by mother's **veins***. Placenta also produces **progesterone***.

Mother's blood vessels

Intervillous spaces. Filled with blood from mother's blood vessels.

Chorionic villi. Finger-like projections containing blood vessels leading from **umbilical cord**.

Layer of tissue called **chorion**

Earlier on, there is a gap between the **amnion** and **chorion**.

Foetus (eight months)

Thin layer of tissue called **amnion**

Amniotic sac containing cushioning **amniotic fluid**.

Umbilical cord. Attaches **foetus** to **placenta**. Contains two **arteries*** and one **vein***.

TYPES OF REPRODUCTION

Reproduction is the creation of new life, a process which occurs in all living things. The two main types are **sexual** and **asexual reproduction**, but there is also a special case called **alternation of generations**.

Sexual reproduction

Sexual reproduction is the type of reproduction shown by all flowering plants and most animals. It involves the joining (**fusion**) of two **gametes** (sex cells), one male and one female. This process is called **fertilization**, and is further described on pages 30 (flowering plants), 91 (humans and similar animals) and 48 (other animals). The two gametes each have only half the number of **chromosomes*** (called the **haploid number***) as the plant or animal which produced them. This is achieved by a special kind of cell division (see pages 94-95) and ensures that when the gametes come together, the new individual produced has the correct, original number of chromosomes (called the **diploid number***).

Sexual reproduction in humans

1. **Sperm** fertilizes **ovum** to form **zygote**.
2. Cell divides in two by **mitosis***.
3. Cells divide again.

1.

2.

3.

4.

4. Cells continue to divide, to form a **morula**, then a **blastocyst** which becomes embedded in the wall of the **uterus***.

Gametes or germ cells
The sex cells which join in **sexual reproduction** to form a new living thing. They are made by a special kind of cell division (see pages 94-95). In animals and simple plants, male gametes are known as **sperm**, short for **spermatozoa** (sing. **spermatozoon**) in animals and **spermatozooids** in simple plants. In flowering plants, they are just **nuclei*** (rather than cells) and are called **male nuclei** (see also pages 30 and 95). Female gametes are called **ova** (sing. **ovum**) or **egg cells** (egg cell is usually used in the case of plants). A sperm is smaller than an ovum and has a "tail" (**flagellum***).

Zygote
The first cell of a new living thing. It is formed when a male and female **gamete** join (see **sexual reproduction**).

Embryo
A new developing individual. It grows from one cell (the **zygote**) by a series of cell divisions (see pages 12-13) called **cleavage**. In humans, this first produces a ball of cells (**morula**) from the one original, and then a larger, hollow ball (**blastocyst**). After **implantation***, this is called the embryo. As it grows, the cells become **differentiated**, i.e. each develops into one kind of cell, e.g. a nerve cell.

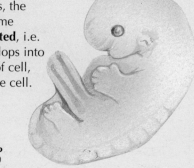

Human **embryo**
(at eight weeks)

* **Chromosomes**, 96; **Diploid number**, 12 (**Mitosis**); **Flagella**, 40; **Haploid number**, 94 (**Meiosis**); **Implantation**, 91 (**Fertilization**); **Nucleus**, 10; **Uterus**, 89.

Asexual reproduction

Asexual reproduction is the simplest form of reproduction, occurring in many simple plants and animals. There are a number of different types, e.g. **binary fission** (a simple organism dividing into two identical ones), **vegetative reproduction***, **gemmation** and **sporulation**, but they all share two main features. Firstly, only one parent is needed and secondly, the new individual is always genetically identical to this parent.

Gemmation

Called **budding** in animals. A type of **asexual reproduction** occurring in many simple plants and animals, e.g. hydra. It involves the formation of a group of cells which grows out of the organism and develops into a new individual. It either breaks away from the parent or (in **colonial*** animals, e.g. corals) it stays attached (though self-contained).

Sporulation

The production of bodies called **spores** by simple plants, e.g. fungi and mosses. After dispersal by wind or water, these develop into new plants. There are two types of spore. One type is produced (e.g. in complex fungi, mosses and ferns) by a special kind of cell division (see pages 94-95) which is a feature of **sexual reproduction**. The new plants are not the same as the parent (see **alternation of generations**). Another kind of spore, however, is produced in plants such as simple fungi by ordinary cell division (see pages 12-13). The spores develop into plants which are identical to the parent (an important feature of **asexual reproduction**). Although only one parent is needed in both cases, true asexual reproduction really only occurs with the second type.

Spores released here

Spores of this common puffball (a complex fungus) disperse through a hole that forms in the ball.

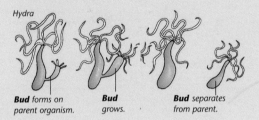

Hydra

Bud forms on parent organism.

Bud grows.

Bud separates from parent.

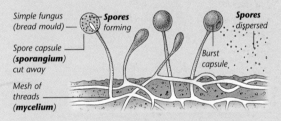

Simple fungus (bread mould)

Spores forming

Spore capsule (**sporangium**) cut away

Mesh of threads (**mycelium**)

Burst capsule

Spores dispersed

Alternation of generations

A reproductive process found in many simple animals and plants, e.g. jellyfish and mosses. In the animals, a form produced by **sexual reproduction** alternates with one produced **asexually**. In the plants, though, the alternation is really between two stages of sexual reproduction. One plant body (**gametophyte**) produces another (**sporophyte**) by sexual reproduction. This then produces **spores** (see **sporulation**) which grow into new gametophytes. However, the spores are made in the same way as **gametes** (see pages 94-95) and they (and the gametophytes) have only half the original number of **chromosomes***. The gametophytes produce gametes by ordinary cell division (see pages 12-13), as there is no need to halve the chromosomes again.

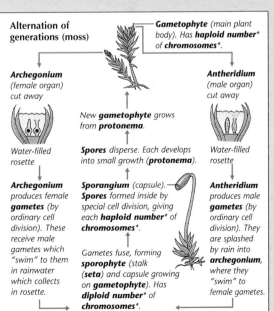

Alternation of generations (moss)

Gametophyte (main plant body). Has **haploid number*** of **chromosomes***.

Archegonium (female organ) cut away

Antheridium (male organ) cut away

*New **gametophyte** grows from **protonema**.*

Water-filled rosette

*Spores disperse. Each develops into small growth (**protonema**).*

Water-filled rosette

Archegonium produces female **gametes** (by ordinary cell division). These receive male gametes which "swim" to them in rainwater which collects in rosette.

Sporangium (capsule). **Spores** formed inside by special cell division, giving each **haploid number*** of **chromosomes***.

*Gametes fuse, forming **sporophyte** (stalk (**seta**) and capsule growing on **gametophyte**). Has **diploid number*** of **chromosomes***.*

Antheridium produces male **gametes** (by ordinary cell division). They are splashed by rain into **archegonium**, where they "swim" to female gametes.

* **Chromosomes**, 96; **Colonial**, 114; **Diploid number**, 12 (**Mitosis**); **Haploid number**, 94 (**Meiosis**); **Vegetative reproduction**, 35.

93

CELL DIVISION FOR REPRODUCTION

Many cells within a living thing can divide to produce new cells for growth and repair (see pages 12-13). There is, however, a second type of cell division, which happens specifically to produce the **gametes*** (sex cells) needed for **sexual reproduction*** (and also one of the two types of **spore***). The division of the **nucleus*** in this type of cell division is called **meiosis**. The production of gametes, including both the cell division and the subsequent maturing of the gametes, is called **gametogenesis**.

Meiosis

The division of the **nucleus*** when a cell divides to produce sex cells (see introduction). It can be split into two separate divisions – the **first meiotic division** (or **reduction division**) and the **second meiotic division** (each is followed by division of the **cytoplasm***). These can be divided into different phases (as in **mitosis***). Meiosis in general, and the first meiotic division in particular, ensures that each new **daughter nucleus** receives exactly half the number of **chromosomes*** as the original nucleus. The original number is the **diploid number** (see **mitosis**, page 12); the halved amount is the **haploid number**.

First meiotic division

These pictures show an animal cell, but only four **chromosomes*** are shown.

Prophase (early stage)

*Threads of **chromatin*** in **nucleus*** coil up to form **chromosomes***. Paired chromosomes (**homologous chromosomes**) line up side by side, forming pairs called **bivalents**. Each chromosome duplicates, becoming a pair of **chromatids** (each group of four chromatids now called a **tetrad**). **Centrioles*** move to opposite poles of cell.*

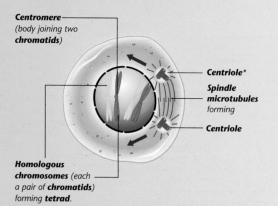

Centromere
(body joining two **chromatids**)

Centriole*

Spindle microtubules forming

Centriole

Homologous chromosomes (each a pair of **chromatids**) forming **tetrad**.

Crossing over (occurs in early prophase)

Chromatids** of each **tetrad** cross over each other at places called **chiasmata** (sing. **chiasma**). Two chromatid pieces (one from each pair) break off and swap over. Causes mixing of **genes (helping to ensure new living things are never identical to parents, i.e. always a new variety of types).*

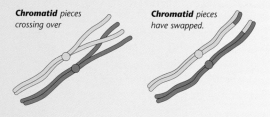

Chromatid pieces crossing over

Chromatid pieces have swapped.

Prophase (later stage)

***Homologous chromosomes** (each a pair of **chromatids**) move together to equator of cell.*

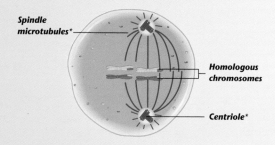

Spindle microtubules*

Homologous chromosomes

Centriole*

Metaphase

Nuclear membrane disappears, two **centrioles*** form a **spindle** (see **metaphase** of **mitosis**, page 13). **Chromosomes*** (pairs of **chromatids**) become attached to spindle by **centromeres**.*

Spindle (made up of microtubules)

Homologous chromosomes

Centromere attached to spindle microtubule

Centriole*

* **Centrioles**, 12; **Chromatin**, 10 (**Nucleus**); **Chromosomes**, 96; **Cytoplasm**, 10; **Gametes**, 92; **Genes**, 97; **Mitosis**, 12; **Nuclear membrane**, 10 (**Nucleus**); **Sexual reproduction**, 92; **Spindle microtubules**, 13; **Spores**, 93 (**Sporulation**).

Anaphase

Homologous chromosomes (each still a pair of **chromatids**) separate (see **Law of segregation**, page 98), dragged apart by **spindle microtubules**.

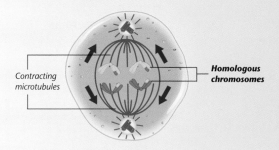

Contracting microtubules

Homologous chromosomes

Telophase

Spindle disappears. Happens in conjunction with **cytokinesis** (division of **cytoplasm***). Two new cells formed, each with half the original number of **chromosomes*** (each two **chromatids**). Brief **interphase*** (intervening period) usually follows, in which case **nuclear membranes*** form and chromosomes uncoil again to form thread-like mass (**chromatin***).

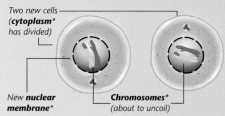

Two new cells (**cytoplasm*** has divided)

New **nuclear membrane***

Chromosomes* (about to uncoil)

Second meiotic division

The **second meiotic division** happens in the cells produced by the **first meiotic division**. It occurs in exactly the same way, and with the same phases, as **mitosis*** (when the **nucleus*** divides as part of cell division for growth and repair) and is followed in the same way by the division of the **cytoplasm***.

The only difference is that each dividing nucleus now has only the **haploid number** of chromosomes* (see **meiosis**) so the resulting new sex cells (**gametes***) will also be haploid. The second division differs according to whether male or female gametes are to be produced, and the final maturing of the gametes after the second division is different in animals and plants (see text below).

Gamete production (male)

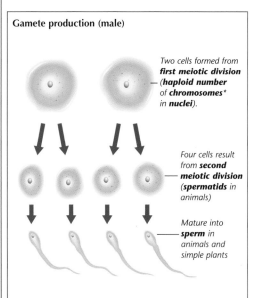

Two cells formed from **first meiotic division** (**haploid number** of **chromosomes*** in **nuclei**).

Four cells result from **second meiotic division** (**spermatids** in animals)

Mature into **sperm** in animals and simple plants

Two cells formed from **first meiotic division** divide again (see **second meiotic division**). In animals, resulting four cells called **spermatids** and mature into male **gametes*** (sex cells), or **sperm**. In simple plants, four cells either develop into sperm or into type of **spore*** involved in **alternation of generations***. In flowering plants, **nuclei*** of four cells each divide again (**mitosis***). Resulting cells (**pollen*** grains) each have two nuclei (one later divides again to form two **male nuclei***).

Gamete production (female)

Two cells formed from **first meiotic division** (**haploid number** of **chromosomes*** in **nuclei**)

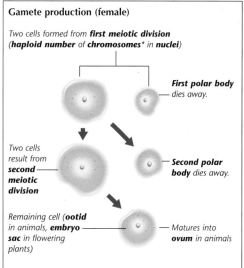

First polar body dies away.

Two cells result from **second meiotic division**

Second polar body dies away.

Remaining cell (**ootid** in animals, **embryo sac** in flowering plants)

Matures into **ovum** in animals

One of two cells formed by **first meiotic division** dies away (called **first polar body**). Other divides again (see **meiotic division**). Of two resulting cells, one (**second polar body**) dies away. In animals, other one called **ootid** and matures into female **gamete*** (sex cell), or **ovum**. In flowering plants, other one called **embryo sac** and its **nucleus*** divides three more times (by **mitosis***). Of eight new nuclei, six have cells form around them, two stay naked. One of six cells is female gamete, or **egg cell** (see **ovule**, page 30). Formation of egg cell in simple plants is very similar.

---

* **Alternation of generations**, 93; **Centrioles**, 12; **Chromatin**, 10 (**Nucleus**); **Chromosomes**, 96; **Cytoplasm**, 10; **Gametes**, 92; **Interphase**, 12; **Male nuclei**, 92 (**Gametes**); **Mitosis**, 12; **Nuclear membrane**, 10 (**Nucleus**); **Pollen**, 30; **Spores**, 93 (**Sporulation**).

95

GENETICS AND HEREDITY

Genetics is a branch of biology. It is the study of **inheritance** – the passing of characteristics from one generation to the next. The bodies which are instrumental in this process are called **chromosomes**. Each chromosome is made up of **genes** – the "coded" instructions for the appearance and constituents of an organism. For more about inheritance, see page 98.

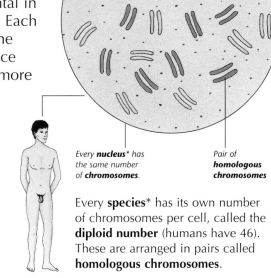

Chromosomes

Structures present at all times in the **nuclei*** of all cells, though they only become independently visible (as thread-like bodies of differing shapes and sizes) when a cell is dividing (and has been stained with a dye). Each one is made with a single molecule of **DNA** (see **nucleic acids**, below), plus proteins called **histones**. The DNA molecule is a chain of many connected **genes**.

*Every **nucleus*** has the same number of **chromosomes**.*

*Pair of **homologous chromosomes***

Every **species*** has its own number of chromosomes per cell, called the **diploid number** (humans have 46). These are arranged in pairs called **homologous chromosomes**.

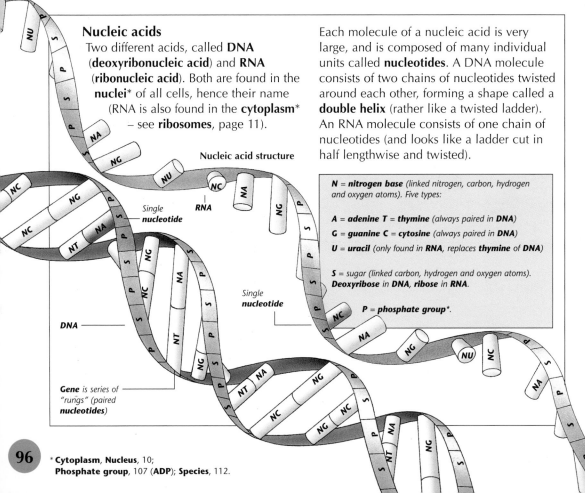

Nucleic acids

Two different acids, called **DNA** (**deoxyribonucleic acid**) and **RNA** (**ribonucleic acid**). Both are found in the **nuclei*** of all cells, hence their name (RNA is also found in the **cytoplasm*** – see **ribosomes**, page 11).

Each molecule of a nucleic acid is very large, and is composed of many individual units called **nucleotides**. A DNA molecule consists of two chains of nucleotides twisted around each other, forming a shape called a **double helix** (rather like a twisted ladder). An RNA molecule consists of one chain of nucleotides (and looks like a ladder cut in half lengthwise and twisted).

Nucleic acid structure

Single nucleotide

RNA

Single nucleotide

DNA

Gene is series of "rungs" (paired **nucleotides**)

N = **nitrogen base** (*linked nitrogen, carbon, hydrogen and oxygen atoms*). *Five types:*

A = **adenine T** = **thymine** (*always paired in **DNA***)

G = **guanine C** = **cytosine** (*always paired in **DNA***)

U = **uracil** (*only found in **RNA**, replaces **thymine** of **DNA***)

S = *sugar* (*linked carbon, hydrogen and oxygen atoms*). **Deoxyribose** in **DNA**, **ribose** in **RNA**.

P = **phosphate group***.

* **Cytoplasm, Nucleus**, 10; **Phosphate group**, 107 (**ADP**); **Species**, 112.

Genes

Sets of "coded" instructions which make up the **DNA** molecule of a **chromosome** (in humans, each DNA molecule is thought to contain about 1,000 genes). Each gene is a connected series of about 250 "rungs" on the DNA "ladder". Since the order of the "rungs" varies, each gene has a different "code", relating to one specific characteristic (**trait**) of the organism, e.g. its **blood group*** or the composition of a **hormone***. With the exception of the **sex chromosomes**, the genes carried on paired **homologous chromosomes** (see **chromosomes**) are also paired, and run down the chromosomes in the same order (one member of each pair on each chromosome). These paired genes control the same characteristic and may give identical instructions. However, their instructions may also be different, in which case the instructions from one gene (the **dominant** gene) will "mask out" those from the other (the **recessive** gene), unless **incomplete dominance** or **codominance** is shown. Two such non-identical genes are called **alleles** or **allelomorphs**.

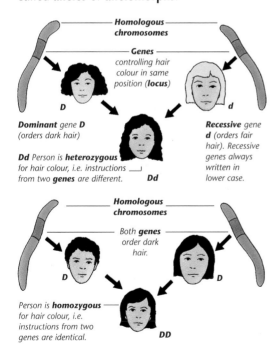

Homologous chromosomes

Genes controlling hair colour in same position (**locus**)

Dominant gene **D** (orders dark hair)

Recessive gene **d** (orders fair hair). Recessive genes always written in lower case.

Dd Person is **heterozygous** for hair colour, i.e. instructions from two **genes** are different. **Dd**

Homologous chromosomes

Both **genes** order dark hair.

Person is **homozygous** for hair colour, i.e. instructions from two genes are identical. **DD**

*The two examples have different **genotypes** for hair colour, i.e. different sets of instructions (**DD** and **Dd**), but are the same **phenotype**, i.e. the resulting characteristic is the same (dark hair).*

Incomplete dominance or blending

A situation where a pair of **genes** which control the same characteristic give different instructions, but neither is **dominant** (see **genes**) or obvious in the result. For example, a lack of dominance between a gene for red colour and one for white results in the intermediate roan colour of some cows.

Incomplete dominance can be used by gardeners to produce flowers of the same species but with a variety of colours. They do this by cross-pollinating flowers of different colours.

White camellia   Red camellia   Pink camellia

Codominance

A special situation where a pair of **genes** controlling the same characteristic give different instructions, neither is **dominant** (see **genes**), but both are represented in the result. The human **blood group*** AB, for example, results from equal dominance between a gene for group A and one for group B.

Sex chromosomes

One pair of **homologous chromosomes** (see **chromosomes**) in all cells (all the others are called **autosomes**). There are two different kinds of sex chromosomes, called the **X** and **Y** **chromosomes**. A male has one X and one Y. The Y chromosome carries the genetic factor (not a **gene** as such) determining maleness, thus all individuals with two X chromosomes are female.

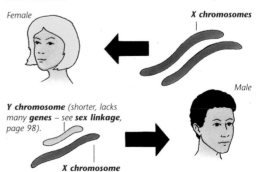

Female

X chromosomes

Male

Y chromosome (shorter, lacks many **genes** – see **sex linkage**, page 98).

X chromosome

Inheriting genes

Every new organism inherits its **chromosomes*** (and **genes***) from its parents. In **sexual reproduction***, the **sperm*** and **ovum*** (sex cells) which come together to form this new individual have only half the normal number of chromosomes (the **haploid number** – see pages 94-95). This ensures that the **zygote*** (first new cell) formed from the two sex cells will have the normal number (see **chromosomes**, page 96). Two laws (**Mendel's laws**) point out genetic factors which are always true when cells divide to produce sex cells.

Law of segregation (Mendel's first law)
Homologous chromosomes* always separate when the **nucleus*** of a cell divides to produce **gametes*** (sex cells – see pages 94-95), hence so too do the paired **genes*** which control the same characteristic. The offspring thus always have paired genes (one member of each pair coming from each parent).

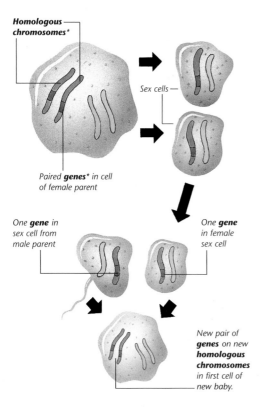

Homologous chromosomes*

Sex cells

Paired **genes*** in cell of female parent

One **gene** in sex cell from male parent

One **gene** in female sex cell

New pair of **genes** on new homologous chromosomes in first cell of new baby.

Law of independent assortment (Mendel's second law)
Each member of a pair of **genes*** can join with either of the two members of another pair when a cell divides to form **gametes*** (sex cells). Hence all the different mixes are possible in a new individual.

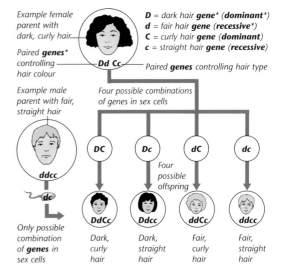

Example female parent with dark, curly hair

Paired **genes*** controlling hair colour

D = dark hair **gene*** (**dominant***)
d = fair hair **gene** (**recessive***)
C = curly hair **gene** (**dominant**)
c = straight hair **gene** (**recessive**)

Dd Cc

Paired **genes** controlling hair type

Example male parent with fair, straight hair

ddcc

dc

Only possible combination of **genes** in sex cells

Four possible combinations of genes in sex cells

DC   **Dc**   **dC**   **dc**

Four possible offspring

DdCc   **Ddcc**   **ddCc**   **ddcc**

Dark, curly hair

Dark, straight hair

Fair, curly hair

Fair, straight hair

Sex linkage
The two **sex (X) chromosomes*** in a female contain many paired **genes*** (like all **chromosomes***), but the **Y chromosome*** in a male lacks partners for most of the genes on its mate (the X). Thus any **recessive*** genes on the X will show up more often in males (see below). The unpaired genes on the X are called **sex-linked genes**.

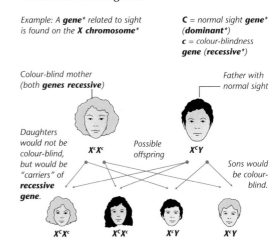

Example: A **gene*** related to sight is found on the **X chromosome***

C = normal sight **gene*** (**dominant***)
c = colour-blindness **gene** (**recessive***)

Colour-blind mother (both **genes recessive**)

Father with normal sight

Daughters would not be colour-blind, but would be "carriers" of **recessive gene**.

X^cX^c

Possible offspring

X^cY

Sons would be colour-blind.

X^cX^c X^cX^c X^cY X^cY

* **Dominant**, 97 (**Genes**); **Homologous chromosomes**, 96 (**Chromosomes**); **Nucleus**, 10; **Ovum**, 92 (**Gametes**); **Recessive**, 97 (**Genes**); **Sexual reproduction**, 92; **Sperm**, 92 (**Gametes**); **X** and **Y chromosomes**, 97 (**Sex chromosomes**); **Zygote**, 92.

GENETIC ENGINEERING

Genetic engineering is the deliberate alteration of **DNA*** within a cell **nucleus*** in order to modify an organism or population of organisms. It is used to create new products which are beneficial to science, agriculture, medicine and industry. New uses for genetically engineered organisms are being discovered all the time.

Gene cloning

The main technique of genetic engineering. Desirable **genes*** are duplicated artificially by inserting DNA molecules (containing the genes) into other organisms, such as fast-breeding bacteria, which then reproduce the DNA. Gene cloning is a complex process. The most common method is shown below.

Gene cloning

*1. DNA containing a particular desirable **gene***, known as **target DNA**, is removed from a donor cell.*

DNA

Donor cell

Cell debris **Target DNA**

*2. Some bacteria contain a **plasmid** – a ring of DNA separate from the bacteria's **chromosomes***. Plasmids are capable of inserting themselves into other organisms. Some also give resistance to particular **antibiotics** (drugs that destroy bacteria). Plasmids can be obtained by breaking up bacteria that contain them.*

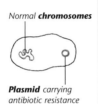

Normal **chromosomes**

Plasmid *carrying antibiotic resistance*

*3. The strands of **plasmid** DNA and **target DNA** are treated so that the ends are sticky. When placed together and heated, the plasmid DNA and target DNA join together. This is called **gene splicing**. The new DNA is known as **recombinant DNA**.*

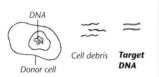

Sticky ends

4. The new plasmids insert themselves into new bacteria not resistant to the antibiotic.

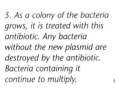

5. As a colony of the bacteria grows, it is treated with this antibiotic. Any bacteria without the new plasmid are destroyed by the antibiotic. Bacteria containing it continue to multiply.

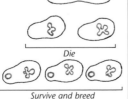

Die

Survive and breed

6. The resultant colony is now made up exclusively of bacteria carrying the antibiotic-resistant plasmids with the target DNA (containing the desired gene). This colony can now be multiplied many times, producing an enormous quantity of the gene.

Uses of genetic engineering

Pharming

The use of plants or animals to produce genetically modified pharmaceutical products. For example, a sheep has been genetically engineered to produce milk which contains **alpha-1 antitrypsin**, a drug which is beneficial to cystic fibrosis patients.

Protein manufacture

The production, in specially-created bacteria "factories", of medically useful proteins such as **insulin*** to help diabetics, and **antihaemophilic globulin** to treat people with haemophilia.

Genetically engineered crops

Plants which have been bred with a greater resistance to disease, pesticides and weather, by inserting foreign **genes*** into their **nuclei***. An example of this technique is shown below.

Some flounders have an antifreeze chemical in their blood, which helps them to survive in freezing water.

*The **gene*** which gives their blood this property can be extracted and introduced into tomato plants. Tomatoes from such plants are now more able to withstand frost and snow. This makes them more readily available at the extremes of their normal growing season.*

* **Chromosomes**, 96; **DNA**, 96 (**Nucleic acids**); **Genes**, 97; **Insulin**, 108; **Nucleus**, 10.

Uses of genetic engineering - continued

Animal cloning
Producing a genetically identical duplicate, or **clone**, of an animal. In 1997, scientists took a cell from a female sheep and placed its **chromosomes*** into another sheep's **ovum***, which had had its own chromosomes removed.

The ovum was planted in the second sheep's womb, and five months later, a lamb, known to the world as Dolly, was born. This experiment proved it was possible to produce a complex living organism, without any kind of **sexual reproduction***.

Sheep created by sexual reproduction*

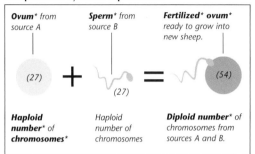

Ovum* from source A

Sperm* from source B

Fertilized* ovum* ready to grow into new sheep.

(27)   **+**   (27)   **=**   (54)

Haploid number* of **chromosomes***

Haploid number of chromosomes

Diploid number* of chromosomes from sources A and B.

Sheep created by gene cloning*

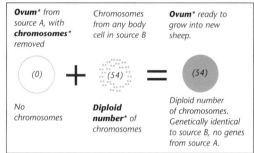

Ovum* from source A, with **chromosomes*** removed

Chromosomes from any body cell in source B

Ovum* ready to grow into new sheep.

(0)   **+**   (54)   **=**   (54)

No chromosomes

Diploid number* of chromosomes

Diploid number of chromosomes. Genetically identical to source B, no genes from source A.

Genetics and the future

Human cloning
The theoretical creation of human life, using the same method that created Dolly the sheep. Genetic engineering techniques are now so advanced that any one of a human body's 100 million million cells could be used to create a new human. But, just as appearance, character and intellect are slightly different in identical twins, a **clone** of another human being would not be a replica, only a person with identical **genes***. A clone would also be a generation apart in age. Possible uses of this technology include the treatment of **infertile** couples – couples who are not naturally able to have children.

Genome mapping
Making a detailed list of the **nucleotides*** contained in the **genome** (genetic code) of any organism. Scientists have already mapped the genome of a yeast cell, and are currently mapping the three billion nucleotides contained in a human genome. They intend to complete this task early in the twenty-first century. The resulting map will enable them to identify every **gene*** in human **chromosomes*** and understand what each one does.

Genetic diagnosis
The identification of illness by examination of **genes***. Scientists can already identify some genetic disorders, which show up as irregularities in the **nucleotide*** sequence. For example, **Huntingdon's chorea** (an illness which causes gradual physical and mental deterioration) can now be detected in a **foetus***. Such research could also make it possible to identify a gene which makes people more susceptible to some cancers. Once identified, treatment could be applied to prevent the cancer from developing.

Organ modification
Introducing genes which encourage body organs to heal themselves. One new technique encourages the hearts of patients needing bypass surgery to grow new blood vessels themselves.

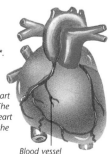

When the heart is growing inside an **embryo***, a particular **gene*** instructs it to construct its **arteries***. The instructions cease once the heart is fully grown. Scientists are developing a method of re-introducing this gene into the heart of a patient with blocked arteries. The gene would enable the damaged heart to grow new blood vessels around the constricted artery, so removing the need for a major heart operation.   Blood vessel

* **Arteries**, 60; **Chromosomes**, 96; **Diploid number**, 12 (**Mitosis**); **Embryo**, 92; **Fertilization**, 91; **Foetus**, 91 (**Pregnancy**); **Genes**, 97; **Gene cloning**, 99; **Haploid number**, 94 (**Meiosis**); **Nucleotides**, 96 (**Nucleic acids**); **Ovum**, 92 (**Gametes**); **Sexual reproduction**, 92; **Sperm**, 92 (**Gametes**).

FLUID MOVEMENT

The movement of substances around the body, especially their movement in and out of cells, is essential to the life of an organism. Food matter must be able to pass into the cells, and waste and harmful material must be able to move out. Most solids and liquids travel round the body in **solutions**, i.e. they (**solutes**) are dissolved in a fluid (the **solvent** – normally water).

Diffusion

The movement of molecules of a substance from an area where they are in higher concentration to one where their concentration is lower. This is a two-way process (where the concentration of a **solute** is low, that of the **solvent** will be high, so its molecules will move the other way) and it ceases when the molecules are evenly distributed. Many substances, e.g. oxygen and carbon dioxide, diffuse into and out of cells.

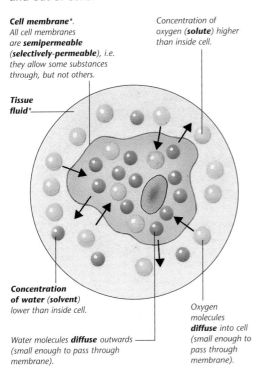

Cell membrane*.
All cell membranes are **semipermeable** (**selectively-permeable**), i.e. they allow some substances through, but not others.

Tissue fluid*

Concentration of oxygen (**solute**) higher than inside cell.

Concentration of water (solvent) lower than inside cell.

Water molecules **diffuse** outwards (small enough to pass through membrane).

Oxygen molecules **diffuse** into cell (small enough to pass through membrane).

Osmosis

The movement of molecules of a **solvent** through a **semipermeable** membrane (see below, left) which lowers the concentration of a **solute** on the other side of the membrane, and evens out the concentrations either side. This is a one-way type of **diffusion**, occurring when the molecules of the solute cannot pass the other way. **Osmotic pressure** is the pressure which builds up in an enclosed space, e.g. a cell, when a **solvent** enters by osmosis.

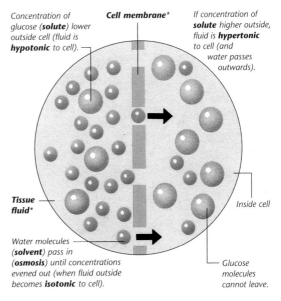

Concentration of glucose (**solute**) lower outside cell (fluid is **hypotonic** to cell).

Cell membrane*

If concentration of **solute** higher outside, fluid is **hypertonic** to cell (and water passes outwards).

Tissue fluid*

Inside cell

Water molecules (**solvent**) pass in (**osmosis**) until concentrations evened out (when fluid outside becomes **isotonic** to cell).

Glucose molecules cannot leave.

Active transport

A process which occurs when substances have to be "pumped" in the opposite direction to that in which they would travel by **diffusion** (i.e. from low to high concentration), e.g. when cells take in large amounts of glucose for breakdown. It is not yet fully understood, but it is thought that special "carrier" molecules within the **cell membrane*** pick up molecules of solute, carry them through the membrane and release them. Energy is needed for this action (since it opposes the natural tendency). This is supplied in the form of **ATP***.

Pinocytosis

The taking in of a fluid droplet by inward-folding and separation of a section of **cell membrane*** (forming a **vacuole***). Most cells can do this.

* **ATP**, 107; **Cell membrane**, 10; **Tissue fluid**, 64; **Vacuoles**, 10.

FOOD AND HOW IT IS USED

Food is vital to all organisms, providing all the materials needed to be broken down for energy, to regulate cellular activities and to build and repair tissues (see pages 104-107). Of the various food substances, **carbohydrates**, **proteins** and **fats** are called **nutrients**, and **minerals**, **vitamins** (not needed by plants) and water are **accessory foods**. Plants build their own nutrients (by **photosynthesis***), and take in minerals and water; animals take in all the substances they need and break them down by digestion (see pages 110-111).

Carbohydrates

A group of substances made up of carbon, hydrogen and oxygen, which exist in varying degrees of complexity (see "terms used", page 111). In animals, complex carbohydrates are taken in and broken down by digestion (see page 110) into the simple carbohydrate **glucose**. The breakdown of glucose (**internal respiration***) provides almost all the energy for life's activities. Plants build up glucose from other substances (see **photosynthesis**, page 26).

Proteins

A group of substances made up of simpler units called **amino acids**. These contain carbon, hydrogen, oxygen, nitrogen and, in some cases, sulphur. Most protein molecules consist of hundreds, maybe thousands, of amino acids, joined together by links called **peptide links** into one or more chains called **polypeptides***. The many different types of protein each have a different arrangement of amino acids. They include the **structural proteins** (the basic components of new cells) and **catalytic proteins** (**enzymes***), which play a vital role in controlling cell processes.

Plants build up amino acids from the substances they take in (by **photosynthesis***), and then build proteins from these amino acids. Animals take in proteins and break them down into single amino acid molecules by digestion (see page 110). These are then transported in the blood to all the body cells and reassembled into the different proteins needed (see **ribosomes**, page 11).

Fats

A group of substances made up of carbon, hydrogen and a small amount of oxygen. Plants build fats from the substances they take in, and their seeds hold most of them as a store of food. This can be converted to extra **glucose** (see **carbohydrates**) to provide energy for the growing plant. Digestion of fats in animals produces **fatty acids** and **glycerol** (see page 110). If these need to be broken down for energy (as well as glucose), this occurs in the liver. This results in some products which the liver can convert to glucose, but others it cannot. These are instead converted elsewhere to a substance which forms a later stage of glucose breakdown.

Fatty acids and glycerol not needed for energy are immediately recombined to form fat particles and stored in various body areas, e.g. under the skin (see **subcutaneous layer**, page 82).

*Like all animals, humans cannot build thier own **nutrients**, and rely on the food they eat for energy. This comes either from plants, e.g. fruit and vegetables (see picture, left), or animals, e.g. meat and milk.*

***Enzymes**, 105; **Internal respiration**, 106; **Photosynthesis**, 26; **Polypeptides**, 111.

Roughage or fibre

Bulky, fibrous food, e.g. bran, and pulses such as lentils and beans. Much of it is made up of **cellulose**, a **carbohydrate** found in plant **cell walls***. Unlike most carbohydrates, cellulose cannot be digested by most animals, including humans, because they lack the necessary **digestive enzyme***, called **cellulase**. (Some animals, e.g. snails, do have this enzyme, and others, like cows, who must digest cellulose, do so in another way – see **rumen**, page 43.) The fact that roughage is bulky and coarse means that food can be gripped by intestinal muscles, and so moved on through the digestive system.

Vitamins

A group of substances vital to animals, though only needed in tiny amounts. The most important function of many vitamins is to act as **co-enzymes***, i.e. to help **enzymes*** catalyse chemical reactions. See page 111 for a list of vitamins and their functions.

Minerals

Natural inorganic substances, e.g. phosphorus and calcium. They form a vital part of plant and animal tissue, e.g. in bones and teeth. Many are found in **enzymes*** and **vitamins**. They include **trace elements**, e.g. copper and iodine, present in tiny amounts.

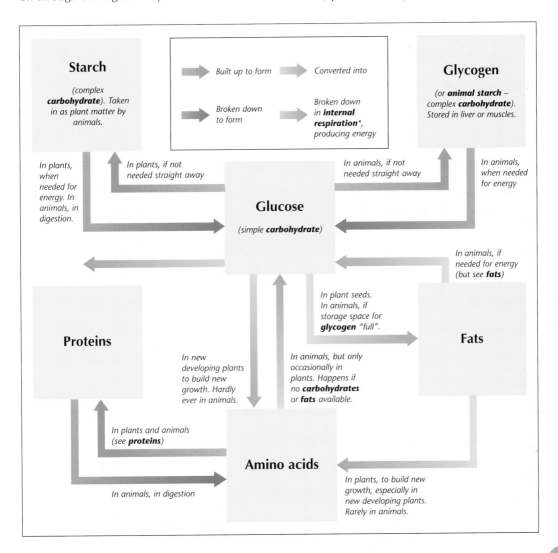

Starch
*(complex **carbohydrate**). Taken in as plant matter by animals.*

Built up to form → Converted into

Broken down to form → Broken down in **internal respiration***, producing energy

Glycogen
*(or **animal starch** – complex **carbohydrate**). Stored in liver or muscles.*

In plants, when needed for energy. In animals, in digestion.

In plants, if not needed straight away

In animals, if not needed straight away

In animals, when needed for energy

Glucose
*(simple **carbohydrate**)*

*In animals, if needed for energy (but see **fats**)*

*In plant seeds. In animals, if storage space for **glycogen** "full".*

Proteins

In new developing plants to build new growth. Hardly ever in animals.

*In animals, but only occasionally in plants. Happens if no **carbohydrates** or **fats** available.*

Fats

*In plants and animals (see **proteins**)*

Amino acids

In plants, to build new growth, especially in new developing plants. Rarely in animals.

In animals, in digestion

* **Cell wall**, 10; **Co-enzymes**, 105 (**Enzymes**);
Digestive enzymes, 110; **Internal respiration**, 106.

103

METABOLISM

Metabolism is a collective term for all the complex, closely-coordinated chemical reactions occurring inside an organism. These can be split into two opposing sets of reactions, called **catabolism** and **anabolism**. The rates of the reactions vary in response to variations in the organism's internal and external environments, and they play a major role in keeping internal conditions stable (see **homeostasis**, page 107).

Catabolism

A collective term for all the reactions which break down substances in the body (**decomposition reactions**). One example is digestion in animals, which breaks down complex substances into simpler ones (see chart, page 110). Another is the further breakdown of these simple substances in the cells (**internal respiration***). Catabolism always liberates energy (in digestion, most is lost as heat, but in internal respiration, it is used for the body's activities). This is despite the fact that, as with all chemical reactions, catabolism itself requires energy. The energy needed is taken from the much greater amount of energy produced during the reactions. The rest of this is released, hence the overall result is always an energy "profit".

*Strenuous exercise, such as cycling, can cause the **metabolic rate** to increase to as much as 15 times the **basal metabolic rate**. The heart rate increases, and more oxygen is taken in. These changes enable food to be **catabolized** more quickly to produce the extra energy needed. One side effect is a rise in body temperature, which prompts the body to produce more sweat.*

Anabolism

A collective term for all the reactions which build up substances in the body (**synthesis reactions**). One example is the linking together of amino acids to form proteins (see page 102). Anabolism always needs energy to be taken in, since the small amount produced during the reactions is never enough (i.e. the overall result of anabolism is an energy "loss"). The extra energy is taken from the **catabolism** "profit".

Metabolic rate

The overall rate at which metabolic reactions occur in an individual. In human beings, it varies widely from person to person, and in the same individual under different conditions. It increases under stress, when the body temperature rises and during exercise, hence the true and accurate measurement of a person's metabolic rate is a measurement taken when the subject is resting, has a normal body temperature, and has not recently exercised. This is called the **basal metabolic rate** (**BMR**) and is expressed in **kilojoules** per square metre of body surface per hour (see measuring method and calculations, opposite).

People with high BMR can eat large amounts without putting on weight, because their **catabolism** of food matter (in the cells) happens so fast that not much fat is stored. This fast rate of reactions also often results in "excess" energy (i.e. energy not needed for **anabolism**), so they may appear to have a lot of "nervous energy". People with low BMR put on weight easily and often appear to have little energy.

The metabolic rate is influenced by a number of **hormones***, especially **STH**, **thyroxin**, **adrenalin** and **noradrenalin**. For more about these, see chart on pages 108-109.

***Hormones**, 108; **Internal respiration**, 106.

Kilojoule

A unit of energy, specifically used in biology when referring to the amount of heat energy produced by the **catabolism** of food, and hence when measuring a person's **basal metabolic rate** (see **metabolic rate**). The calculations involved in measuring BMR combine certain known facts about the number of kilojoules produced by the breakdown of different substances, with a measurement of oxygen consumption obtained under controlled conditions (see below and right).

To work out a person's basal metabolic rate
(BMR = kJ m^{-2} hr^{-1})

1. Facts known (worked out using a piece of apparatus called a **calorimeter***):*
a) If 1 litre of oxygen is used to break down carbohydrates, c.21.21kJ are produced (i.e. enough heat energy to heat c.5,050g water by 1°C).
b) With fats, the result from 1 litre oxygen is c.19.74kJ.
c) With proteins, the result from 1 litre oxygen is c.19.32kJ.

Calculations (example):

1. (Measured) Subject used 1.5 litres oxygen in 5 minutes.

2. Hence he would use 18 litres oxygen in 1 hour (1.5 × 12).

3. (Known) 20.09kJ produced when food broken down by 1 litre oxygen.

4. Hence 361.62kJ produced if food broken down by 18 litres oxygen (20.09 × 18).

5. Hence 361.62kJ produced by breakdown of food in whole of subject's body in 1 hour (he would use 18 litres oxygen per hour – see point 2).

6. But BMR is measured in kJ per square metre of body surface per hour.

7. Standard chart used to work out body surface in square metres.

8. 361.62 divided by body surface (e.g. 2m²) = 180.81kJ m⁻² hr⁻¹ (BMR).

2. First calculation:
Heat energy generated when food (in general) is broken down using 1 litre of oxygen = the average of the three figures above, i.e. 20.09kJ (provided subject measured has taken in equal amounts of all three foodstuffs).

3. Measure the oxygen used by the subject's body in a fixed time. Done using a spirometer (respirometer) – see picture, right.

A trace is drawn as the cylinder moves up and down.

The overall trend of the trace is up (the cylinder moves down as its volume of oxygen decreases).

Drum rotates

Oxygen in cylinder

The air (minus some oxygen) that is breathed out returns to cylinder.

Soda lime absorbs the carbon dioxide.

The subject breathes in from the cylinder this way.

The subject breathes out to the cylinder this way.

Enzymes

Special proteins (**catalytic proteins**) found in all living things and vital to the chemical reactions of life. They act as **catalysts**, i.e. they speed up reactions without themselves being changed. Many enzymes are aided by other substances, called **co-enzymes**, whose molecules are able to "carry" the products of one reaction (catalysed by an enzyme) on to the next reaction.

There are many different types of enzyme, e.g. **digestive enzymes**, which control the breakdown of complex food material into simple soluble substances (for more about these, see text and chart, pages 110-111), and **respiratory enzymes**, which control the further breakdown of these simple substances in the cells to liberate energy (i.e. **internal respiration** – see page 106).

ENERGY FOR LIFE AND HOMEOSTASIS

A living thing needs energy for its activities. This energy comes from a series of chemical reactions inside its cells, known as **internal respiration**, **tissue respiration** or **cellular respiration**. The cells contain various simple food substances, which are the results of digestive breakdown in animals (see pages 110-111) and **photosynthesis*** in plants. These substances all contain stored energy, which is released when internal respiration breaks them down. In almost all cases, glucose is the substance broken down (see **carbohydrates** and diagram, pages 102-103). There are two kinds of respiration – **anaerobic** and **aerobic respiration**.

Anaerobic respiration

A type of **internal respiration** which does not need free oxygen (oxygen taken into the body). It takes place in the cells of all organisms and releases a small amount of energy. In most organisms, it consists of a chain of chemical reactions called **glycolysis**, which break down glucose into **pyruvic acid**. In normal circumstances this is then immediately followed by **aerobic respiration**, which breaks down this poisonous acid in the presence of oxygen. This breakdown releases the bulk of energy. In abnormal conditions, however, it may not be possible for the aerobic stage to follow immediately, in which case a further stage of anaerobic respiration occurs (see **oxygen debt**).

In some microscopic organisms, e.g. yeast and some bacteria, anaerobic respiration always runs through all its stages, providing enough energy for their needs without requiring oxygen.

Aerobic respiration

A type of **internal respiration** which can only take place in the presence of free oxygen (oxygen taken into the body). It is the way in which most living things obtain the bulk of their energy and follows a stage of **anaerobic respiration**. Oxygen (brought by the blood) is taken into each cell and reacts in the **mitochondria*** with the **pyruvic acid** produced in anaerobic respiration. Carbon dioxide and water are the final products of the reactions, and chemical energy is released, which is then "stored" as **ATP**.

Aerobic respiration is an example of **oxidation** – the breakdown of a substance in the presence of oxygen.

Summary of aerobic stages of respiration reactions

$$5O_2 \ + \ 2C_3H_4O_3 \ \rightarrow \ 6CO_2 \ + \ 4H_2O$$

| Oxygen | Pyruvic acid | Carbon dioxide | Water |

Oxygen debt

A situation which occurs when extreme physical exercise is undertaken by an organism which shows **aerobic respiration**. Under these circumstances, the oxygen in the organism's cells is used up faster than it can be taken in. This means that there is not enough to break down the poisonous **pyruvic acid** produced in the first, **anaerobic**, stage of respiration. Instead, the acid undergoes further anaerobic reactions to convert it to **lactic acid** (much less harmful). This begins to build up, and the organism is said to have acquired an oxygen debt. This is "paid off" later by taking in oxygen faster than usual to break down the lactic acid, by breathing heavily, for example.

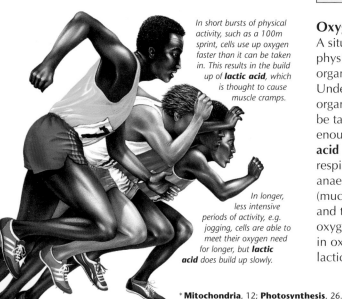

*In short bursts of physical activity, such as a 100m sprint, cells use up oxygen faster than it can be taken in. This results in the build up of **lactic acid**, which is thought to cause muscle cramps.*

*In longer, less intensive periods of activity, e.g. jogging, cells are able to meet their oxygen need for longer, but **lactic acid** does build up slowly.*

*****Mitochondria**, 12; **Photosynthesis**, 26.

ADP (adenosine diphosphate) and ATP (adenosine triphosphate)

Two substances which consist of a chemical grouping called **adenosine**, combined with two and three **phosphate groups** respectively. These phosphate groups each consist of linked phosphorus, oxygen and hydrogen atoms. A phosphate group can combine with other substances (either by itself or linked with other phosphate groups in a chain).

When **aerobic respiration** occurs, the chemical energy released is involved in reactions which result in a conversion of ADP molecules into ATP molecules (by attachment of a third phosphate group in each case). The energy taken in to effect these reactions can be regarded as being "stored" in the form of ATP. This is a substance which can be easily stored in all cells. It is found in especially large quantities in cells which require a lot of energy, e.g. muscle cells. When the energy is needed, reactions occur which convert ATP back into ADP. These reactions result in an overall release of energy – the "stored" energy. In this way, power is supplied for the cell's activities.

ADP and ATP

Adenine + Ribose = Adenosine

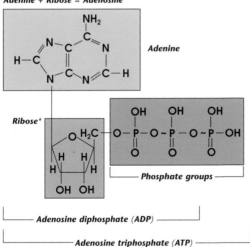

Conversion of ATP into ADP

ATP	⇌	ADP	+	P	+	E
Adenosine triphosphate		Adenosine diphosphate		Phosphate group		Energy

└─ The reaction is reversible.

Homeostasis

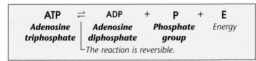

Crocodile gaping to lose heat

Homeostasis is the maintenance, by an organism, of a stable **internal environment**, i.e. a constant temperature, stable composition, level and pressure of body fluids, constant **metabolic rate***, etc. This is vital if the organism is to function properly.

Homeostasis requires the detection of any deviation from the norm (caused by new internal or external factors) and the means to correct such deviations, and is practised most efficiently in birds and **mammals***, e.g. humans. Their detection of deviations is achieved by the **feedback** of information to controlling organs. The blood glucose level, for example, is constantly being detected by the pancreas (i.e. information is "fed back"). The correction of deviations is achieved by **negative feedback**, i.e. feedback which "tells" of deviations and results in a change of action. If the glucose level gets too high, for example, the pancreas reacts by producing more **insulin*** to reduce it (see also **antagonistic hormones**, page 108).

Most homeostatic actions are, like the insulin example, controlled by hormones, many of which are in turn controlled by the **hypothalamus*** in the brain. An example of the importance of the hypothalamus in homeostasis is the control of body temperature. All birds and mammals, e.g. humans, are **homiothermic** (warm-blooded), i.e. they can keep a constant temperature (about 37°C in humans) regardless of external conditions (the opposite is **poikilothermic**, or cold-blooded). A "thermostat" area of the hypothalamus, called the **preoptic area**, detects any changes in body temperature and sends impulses either to the **heat-losing centre** or to the **heat-promoting centre** (both also found in the hypothalamus). These areas then send out nervous impulses to cause various heat-losing or heat-promoting actions in the body.

*Penguins are **homiothermic**. They can generate enough heat to keep themselves and their eggs and young warm.*

*Hypothalamus, 75; Insulin, 108; Mammals, 113; Metabolic rate, 104; Ribose, 96.

107

HORMONES

Hormones are special chemical "messengers" which control various activities inside an organism. These pages deal with the hormones produced by humans and their related groups. Plants also produce hormones (**phytohormones**), though these are not yet fully understood (see **abscission layer**, page 21, and **photoperiodism** and **growth hormones**, page 23).

Human hormones are secreted by **endocrine glands***. Some act only on specific body parts (**target cells** or **target organs**), others cause a more general response. The principal controller of hormone production is the **hypothalamus*** (part of the brain). It controls the secretions of many glands, mainly through its control of the **pituitary gland***, which itself controls many other glands. The hypothalamus "tells" the pituitary to produce its hormones by sending **regulating factors** to its **anterior lobe** and nervous impulses to its **posterior lobe**. Hormone secretion is vital to **homeostasis***.

Regulating factors

Special chemicals which control the production of a number of hormones, and hence many vital body functions. They are sent to the **anterior lobe** of the **pituitary gland*** by the **hypothalamus*** (part of the brain). There are two types – **releasing factors**, which make the gland secrete specific hormones, and **inhibiting factors**, which make it stop its secretion. For example, **FSHRF** (**FSH releasing factor**) and **LHRF** (**LH releasing factor**) cause the release of the hormones **FSH** and **LH** (see chart), and hence the onset of **puberty***. Many regulating factors are vital to **homeostasis***.

Antagonistic hormones

Hormones that produce opposite effects. **Glucagon** and **insulin** (see chart) are examples. When the blood glucose level drops too far, the pancreas produces glucagon to raise it again. A high glucose level causes the pancreas to produce insulin to lower the level (see also **homeostasis**, page 107).

Hormones	
ACTH (**adrenocorticotropic hormone**) or **adrenocorticotropin**	
TSH (**thyroid-stimulating hormone**) or **thyrotropin**	
STH (**somatotropic hormone**) or **somatotropin** or **HGH** (**human growth hormone**)	
FSH (**follicle-stimulating hormone**)	
LH (**luteinizing hormone**). Also called **luteotropin** in women and **ICSH** (**interstitial cell stimulating hormone**) in men.	
Lactogenic hormone or **PR** (**prolactin**)	
Oxytocin	
ADH (**anti-diuretic hormone**) or **vasopressin**	
Thyroxin	
TCT (**thyrocalcitonin**) or **calcitonin**	
PTH (**parathyroid hormone**) or **parathyrin** or **parathormone**	
Adrenalin or **adrenin** or **epinephrin** **Noradrenalin** or **norepinephrin**	
Aldosterone	
Cortisone **Hydrocortisone** or **cortisol**	
Oestrogen (female **sex hormone**) **Progesterone** (female **sex hormone**)	
Androgens (male **sex hormones**), especially **testosterone**	
Gastrin	
CCK (**cholecystokinin**)	
Secretin **PZ** (**pancreozymin**)	
Enterocrinin	
Insulin	
Glucagon	

* **Endocrine glands**, 69; **Homeostasis**, 107; **Hypothalamus**, 75; **Pituitary gland**, 69; **Puberty**, 90.

Where produced	Effects
Pituitary gland *(page 69)* **(anterior lobe)**	*Stimulates production of hormones in* **cortex** *of* **adrenal glands** *(page 69).*
Pituitary gland *(page 69)* **(anterior lobe)**	*Stimulates production of* **thyroxin** *by* **thyroid gland** *(page 69).*
Pituitary gland *(page 69)* **(anterior lobe)**	*Stimulates growth by increasing rate at which amino acids are built up to make proteins in cells.*
Pituitary gland *(page 69)* **(anterior lobe)**	*In women, works with* **LH** *to stimulate development of* **ova** *in* **ovarian follicles** *(page 89) and secretion of* **oestrogen** *by follicles in early stages of* **menstrual cycle** *(page 90). In men, causes formation of* **sperm** *(page 92).*
Pituitary gland *(page 69)* **(anterior lobe)**	*Stimulates* **ovulation** *(page 90), formation of* **corpus luteum** *(page 90) and its secretion of* **oestrogen** *and* **progesterone**. *Works with* **oestrogen** *and* **progesterone** *to stimulate thickening of lining of* **uterus** *(page 89). In men, causes production of* **androgens**.
Pituitary gland *(page 69)* **(anterior lobe)**	*Works with* **LH** *to cause secretion of hormones by* **corpus luteum** *(page 90). Causes milk production after giving birth.*
Hypothalamus *(page 75). Builds up in* **pituitary gland** *(posterior lobe)*	*Stimulates contraction of muscles of* **uterus** *(page 89) during labour and secretion of milk after giving birth.*
Hypothalamus *(page 75). Builds up in* **pituitary gland** *(posterior lobe)*	*Increases amount of water re-absorbed into blood from* **uriniferous tubules** *(page 73) in kidneys.*
Thyroid gland *(page 69)*	*Increases rate of food breakdown, hence increasing energy and raising body temperature. Works with* **STH** *in the young to control rate of growth and development. Contains iodine.*
Thyroid gland *(page 69)*	*Decreases level of calcium and phosphorus in blood by reducing their release from bones (where they are stored).*
Parathyroid glands *(page 69)*	*Increases level of calcium in blood by increasing its release from bone (see above). Decreases phosphorus level.*
Adrenal glands *(page 69)* **(medulla)**. *Also at nerve endings. Secreted at times of excitement or danger.*	*Stimulate liver to release more glucose into blood, to be broken down for energy. Stimulate increase in heart rate, faster breathing and blood vessel constriction.*
Adrenal glands *(page 69)* **(cortex)**	*Increases amount of sodium and water in blood by causing re-absorption of more from* **uriniferous tubules** *(page 73) in kidneys.*
Adrenal glands *(page 69)* **(cortex)**	*Stimulate increase in rate of food breakdown for energy, and thus increase resistance to stress. Lessen inflammation.*
Mostly in **ovarian follicles** *(page 89) and* **corpus luteum** *(page 90) in* **ovaries** *(female sex organs, page 89). Also in* **placenta** *(page 91) during pregnancy.*	*Oestrogen activates development of* **secondary sex characters** *at* **puberty** *(page 90), e.g. breast growth. Both prepare* **mammary** *(milk)* **glands** *for milk production and work with* **LH** *to cause thickening of lining of* **uterus** *(page 89). Progesterone dominates towards end of* **menstrual cycle** *(page 90) and during pregnancy, when it maintains uterus lining and mammary gland readiness.*
Mostly in **interstitial cells** *in* **testes** *(male sex organs, page 88).*	*Activate development and maintenance of* **secondary sex characters** *at* **puberty** *(page 90), e.g. beard growth.*
Cells in stomach	*Stimulates production of* **gastric juice** *(page 110).*
Cells in small intestine	*Stimulates opening of* **sphincter of Oddi**, *contraction of* **gall bladder** *and release of* **bile** *(all page 69) into* **duodenum** *(page 67).*
Cells in small intestine	*Stimulate pancreas to produce* **pancreatic juice** *(page 110) and secrete it into* **duodenum** *(page 67).*
Cells in small intestine	*Stimulates production of* **intestinal juice** *(page 110).*
Pancreas, when blood glucose level too high.	*Stimulates liver to convert more glucose to glycogen for storage (page 103). Also speeds up transport of glucose to cells.*
Pancreas, when blood glucose level too low.	*Stimulates faster conversion of glycogen to glucose in liver (page 103), and conversion of fats and proteins to glucose.*

DIGESTIVE JUICES AND ENZYMES

All the **digestive juices*** of the human body (secreted into the intestines by **digestive glands***) contain **enzymes*** which control the breakdown of food into simple soluble substances. These are called **digestive enzymes** and can be divided into three groups. **Amylases** (or **diastases**) promote the breakdown of **carbohydrates*** (the final result being **monosaccharides** – see terms used, right). **Proteinases** (or **peptidases**) promote the breakdown of **proteins** into **amino acids*** by attacking the **peptide links** (see **proteins**, page 102). **Lipases** promote the breakdown of **fats** into **glycerol** and **fatty acids** (see **fats**, page 102). The chart below lists the different digestive juices of the body, together with their enzymes and the action of these enzymes.

Digestive juice: Saliva

Produced by: **Salivary glands*** in mouth

Digestive enzyme: **Salivary amylase** (or **ptyalin**)

Actions: Starts breakdown of **carbohydrates*** starch and glycogen (**polysaccharides**). See page 103.

Products: Some **dextrin** (shorter **polysaccharide**). See note 1.

Digestive juice: Bile

Produced by: Liver. Stored in **gall bladder***, secreted into small intestine (see **CCK**, page 108).

Constituents: **Bile salts** and **bile acids**

Actions: Break up **fats*** (and intermediate compounds) into smaller particles, a process called **emulsification**.

Digestive juice: Gastric juice

Produced by: **Gastric glands*** in stomach lining. Secreted into stomach (see **gastrin**, page 108).

Digestive enzymes (and one other constituent):
1. **Pepsin** (**proteinase**). See note 2.
2. **Rennin** (**proteinase**). Found only in the young.
3. **Hydrochloric acid**
4. **Gastric lipase**. Found mainly in the young.

Actions:
1. Starts breakdown of **proteins*** (**polypeptides**).
2. Works (with calcium) to curdle milk, i.e. to act on its **protein** (**casein**). See note 3.
3. Activates **pepsin** (see note 2), curdles milk in adults (see note 3) and kills bacteria.
4. Starts breakdown of **fat*** molecules in milk.

Products:
1. Shorter **polypeptides**
2, 3. **Curds**, i.e. milk solids
4. Intermediate compounds

Digestive juice: Intestinal juice (or **succus entericus**)

Produced by: **Intestinal glands*** in small intestine lining. Final secretion into small intestine (see **enterocrinin**, page 108).

Digestive enzymes:
1. **Maltase** (**amylase**)
2. **Sucrase** (or **invertase** or **saccharase**) (**amylase**)
3. **Lactase** (**amylase**)
4. **Enterokinase**. See note 2.

Actions:
1. Breaks down **maltose** (**disaccharide**).
2. Breaks down **sucrose** (**disaccharide**).
3. Breaks down **lactose** (**disaccharide**).
4. Completes breakdown of **proteins*** (**dipeptides**).

Products:
1. **Glucose** (or **dextrose**) (**monosaccharide**)
2. **Glucose** and **fructose** (**monosaccharides**)
3. **Glucose** and **galactose** (**monosaccharides**)
4. **Amino acids***

Digestive juice: Pancreatic juice

Produced by: Pancreas. Secreted into small intestine (see **secretin** and **PZ**, page 108).

Digestive enzymes:
1. **Trypsin** (**proteinase**). See note 2.
2. **Chymotrypsin** (**proteinase**). See note 2.
3. **Carboxypeptidase** (**proteinase**). See note 2.
4. **Pancreatic amylase** (or **amylopsin**)
5. **Pancreatic lipase**

Actions:
1, 2, 3. Continue breakdown of **proteins*** (long and shorter polypeptides).
4. Continues breakdown of **carbohydrates***.
5. Breaks down **fat*** particles.

Products:
1, 2, 3. **Dipeptides** and some **amino acids***.
4. **Maltose** (**disaccharide**)
5. **Glycerol** and **fatty acids** (see **fats**, page 102).

Notes

1. Not much **dextrin** is produced at this stage, since food is not in the mouth long enough. Most carbohydrates pass through unchanged.

2. **Proteinases** are first secreted in inactive forms, to prevent them from digesting the digestive tube (made of **protein***, like most of the body). Once in the tube (beyond a protective layer of **mucous membrane***), these are converted into active forms. **Hydrochloric acid** changes **pepsinogen** (inactive) into **pepsin**, **enterokinase** changes **trypsinogen** into **trypsin**, and trypsin then changes **chymotrypsinogen** and **procarboxypeptidase** into **chymotrypsin** and **carboxypeptidase**.

3. The action of **rennin** and **hydrochloric acid** in curdling milk is vital, since liquid milk would pass through the system too fast to be digested.

Terms used

Polysaccharides

The most complex **carbohydrates***. Each is a chain of **monosaccharide** molecules. Most carbohydrates taken into the body are polysaccharides, e.g. **starch** (the main polysaccharide in edible plants) and **glycogen** (the main one in animal matter). For more about starch and glycogen, see page 103.

Disaccharides

Compounds of two **monosaccharide** molecules, either forming intermediate stages in the breakdown of **polysaccharides** or (in the case of **sucrose** and **lactose**) taken into the body as such. (Sucrose is found in sugar beet and sugar cane, lactose occurs in milk.)

Monosaccharides

The simplest **carbohydrates***. Almost all result from **polysaccharide** breakdown, though **fructose** is taken into the body as such (e.g. in fruit juices), as well as resulting from **sucrose** breakdown. **Glucose** is the final result of all action on carbohydrates (fructose and **galactose** are converted to glucose in the liver).

Polypeptides

The complex form taken by all **proteins** entering the body. Each is a chain of hundreds (or thousands) of **amino acid*** molecules (see **proteins**, pages 102).

Dipeptides

Chains of two **amino acid** molecules, forming intermediate stages in the breakdown of **polypeptides**.

Vitamins and their uses

Vitamin A (retinol)

Sources: Liver, kidneys, fish-liver oils, eggs, dairy products, margarine, **pigment*** (**carotene**) in green and yellow fruit and vegetables, especially tomatoes and carrots (carotene converted to vitamin A in intestines).

Uses: Maintains general health of **epithelial*** cells (lining cells), aids growth, especially bones and teeth. Essential for vision in dim light – involved in formation of light-sensitive **pigment*** (**rhodopsin**), found in **rods** of **retina***. Aids in resistance against infection.

Vitamin B complex

Group of at least 10 vitamins, usually occurring together. Include:
Thiamine (or **aneurin**) (**B1**)
Riboflavin (**B2**)
Niacin (or **nicotinic acid** or **nicotinamide**) (**B3**)
Pantothenic acid (**B5**)
Pyridoxine (**B6**)
Cyanocobalamin (or **cobalamin**) (**B12**)
Folic acid (**Bc** or **M**)
Biotin (sometimes called **vitamin H**)
Lecithin.

Sources: All found in yeast and liver. All except B12 found in wholewheat cereals and bread, wheatgerm and green vegetables, e.g. beans (B12 not found in any vegetable products). B2 and B12 found especially in dairy products. Most also found in eggs, nuts, fish, lean meat, kidneys and potatoes. B6, folic acid and biotin also made by bacteria in intestines.

Uses: Most needed for growth and maintenance of healthy tissues, e.g. muscles (B1, B6), nerves (B1, B3, B6, B12), skin (B2, B3, B5, B6, B12) and hair (B2, B5). Several also aid continuous function of body organs (B5, B6, lecithin). All except folic acid, biotin and lecithin are essential **co-enzymes***, aiding in breakdown of foods for energy (**internal respiration***). Many (especially B2, B6, B12) also co-enzymes aiding build-up of substances (**proteins***) for growth and regulatory or defence purposes. B12 and folic acid vital to formation of blood cells, B5 and B6 vital to manufacture of nerve chemicals (**neurotransmitters***).

Vitamin C (ascorbic acid)

Sources: Green vegetables, potatoes, tomatoes, citrus fruit, e.g. oranges, grapefruit, lemons.

Uses: Needed for growth and maintenance of healthy tissues, especially skin, blood vessels, bones, gums, teeth. Essential **co-enzyme*** in many metabolic reactions, especially **protein*** breakdown and build-up of **amino acids*** into new proteins (especially **collagen** – see **connective tissue**, page 52). Aids in resistance against infection and healing of wounds.

Vitamin D (calciferol)

Sources: Liver, fish-liver oils, oily fish, dairy products, egg yolk, margarine, special substance (provitamin D3) in skin cells (converted to vitamin D when exposed to sunlight).

Uses: Essential for absorption of calcium and phosphorus, and their deposition in bones and teeth. May work with **PTH*** (**hormone**).

Vitamin E (tocopherol)

Sources: Meat, egg yolk, leafy vegetables, nuts, dairy products, margarine, cereals, wholemeal bread, wheatgerm, seeds, seed and vegetable oils.

Uses: Not yet fully understood. Protects membranes from some molecules which could bind and cause cancer.

Vitamin K (phylloquinone or menaquinone)

Sources: Liver, fruit, nuts, cereals, tomatoes, green vegetables, especially cabbage, cauliflower, spinach. Also made by bacteria in intestines.

Uses: Essential for formation of **prothrombin*** in liver (needed to cause clotting of blood).

*** Amino acids**, 102 (**Proteins**); **Carbohydrates**, 102; **Co-enzymes**, 105 (**Enzymes**); **DNA**, 96 (**Nucleic acids**); **Epithelium**, 82 (**Epidermis**); **Internal respiration**, 106; **Neurotransmitters**, 77 (**Synapse**); **Pigments**, 27; **Proteins**, 102; **Prothrombin**, 59 (**Clotting**); **PTH**, 108; **Retina**, 85.

THE CLASSIFICATION OF LIVING THINGS

Callicore cyllene

Callicore mengeli

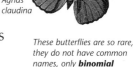

Agrias claudina

Classification, also called **taxonomy**, is the grouping together of living things according to the characteristics they share. The main, formal type of classification (**classical taxonomy**) bases its groups primarily on structural characteristics. The resulting classification charts first list the largest groups – **Kingdoms** – and then list the smaller and smaller sub-divisions within these groups.

*These butterflies are so rare, they do not have common names, only **binomial** names, see below.*

The first groups after the Kingdoms are called **phyla** (sing. **phylum**) in the case of animals, and **divisions** in the case of plants. After these come **classes**, **orders**, **families**, **genera** (sing. **genus**) and finally **species** – the smallest groupings. A species is defined as a group of interbreeding organisms, reproductively isolated from other groups. If this is impossible to establish, then a species is recognized on a **morphological** basis, i.e. by its external appearance.

Some divisions or phyla, especially those with only a few members, may not have all these groups, so the next group after a phylum may be an order, family, genus or even a species. There are also further "mid-way" groups in some cases, such as **sub-kingdoms**, **sub-phyla** or **sub-classes**.

There are areas which are still under dispute in both plant and animal classification. Most scientists recognize five main Kingdoms (see diagram, page 113), but some prefer to group living things into four Kingdoms: **animals** (including **Protista**), **plants** (including **fungi** and **algae**), **Monera** and **viruses**. The diagram below shows the divisions of the Plant Kingdom.

Nomenclature

The naming of organisms. Names of **species** are given in Latin so that all biologists world-wide can follow the same system. This is necessary as species are often known by various common names in different parts of the world. For example, one type of herring, *Alosa pseudoharengus*, has six different names throughout its geographical range.

Every organism has a two word name. This is called the **binomial system**. The first word is a **generic name** (from the **genus**) and the second identifies the species within that genus. The Latin names are governed by The International Commission for Zoological Nomenclature, at the Natural History Museum in London. Most names make direct reference to specific characteristics of the species, such as size, shape or habitat. The giant anteater, for example, is called *Myrmecophaga tridactyla* (*myrmeco* = ant, *phag* = eat, *tri* = three, and *dactyl* = fingers). This describes the food it eats and the three large digging claws at the end of each foreleg. *Giant anteater*

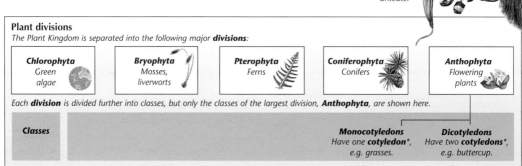

Plant divisions
*The Plant Kingdom is separated into the following major **divisions**:*

Chlorophyta Green algae	**Bryophyta** Mosses, liverworts	**Pterophyta** Ferns	**Coniferophyta** Conifers	**Anthophyta** Flowering plants

*Each **division** is divided further into classes, but only the classes of the largest division, **Anthophyta**, are shown here.*

Classes		**Monocotyledons** Have one **cotyledon***, e.g. grasses.	**Dicotyledons** Have two **cotyledons***, e.g. buttercup.

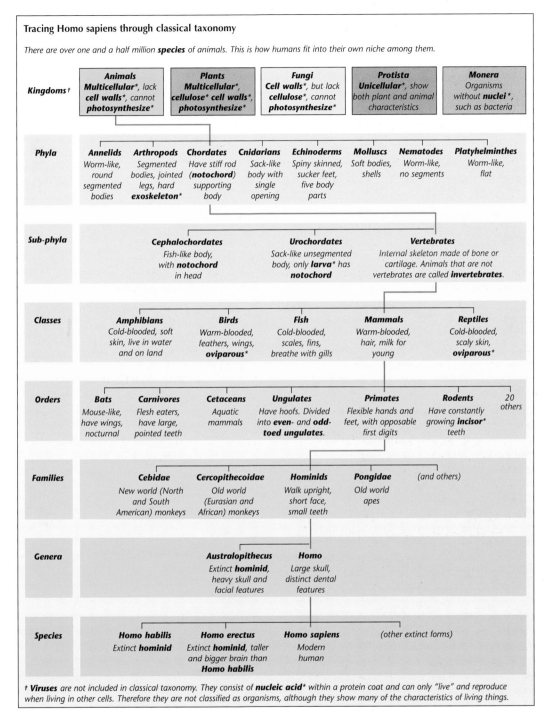

Tracing Homo sapiens through classical taxonomy

There are over one and a half million **species** of animals. This is how humans fit into their own niche among them.

Kingdoms†

Animals	**Plants**	**Fungi**	**Protista**	**Monera**
Multicellular*, lack **cell walls***, cannot **photosynthesize***	**Multicellular***, **cellulose* cell walls***, cannot **photosynthesize***	**Cell walls***, but lack **cellulose***, cannot **photosynthesize***	**Unicellular***, show both plant and animal characteristics	Organisms without **nuclei***, such as bacteria

Phyla

Annelids	**Arthropods**	**Chordates**	**Cnidarians**	**Echinoderms**	**Molluscs**	**Nematodes**	**Platyhelminthes**
Worm-like, round segmented bodies	Segmented bodies, jointed legs, hard **exoskeleton***	Have stiff rod (**notochord**) supporting body	Sack-like body with single opening	Spiny skinned, sucker feet, five body parts	Soft bodies, shells	Worm-like, no segments	Worm-like, flat

Sub-phyla

Cephalochordates	**Urochordates**	**Vertebrates**
Fish-like body, with **notochord** in head	Sack-like unsegmented body, only **larva*** has **notochord**	Internal skeleton made of bone or cartilage. Animals that are not vertebrates are called **invertebrates**.

Classes

Amphibians	**Birds**	**Fish**	**Mammals**	**Reptiles**
Cold-blooded, soft skin, live in water and on land	Warm-blooded, feathers, wings, **oviparous***	Cold-blooded, scales, fins, breathe with gills	Warm-blooded, hair, milk for young	Cold-blooded, scaly skin, **oviparous***

Orders

Bats	**Carnivores**	**Cetaceans**	**Ungulates**	**Primates**	**Rodents**	20 others
Mouse-like, have wings, nocturnal	Flesh eaters, have large, pointed teeth	Aquatic mammals	Have hoofs. Divided into **even-** and **odd-toed ungulates**.	Flexible hands and feet, with opposable first digits	Have constantly growing **incisor*** teeth	

Families

Cebidae	**Cercopithecoidae**	**Hominids**	**Pongidae**	(and others)
New world (North and South American) monkeys	Old world (Eurasian and African) monkeys	Walk upright, short face, small teeth	Old world apes	

Genera

Australopithecus	**Homo**
Extinct **hominid**, heavy skull and facial features	Large skull, distinct dental features

Species

Homo habilis	**Homo erectus**	**Homo sapiens**	(other extinct forms)
Extinct **hominid**	Extinct **hominid**, taller and bigger brain than **Homo habilis**	Modern human	

† **Viruses** are not included in classical taxonomy. They consist of **nucleic acid*** within a protein coat and can only "live" and reproduce when living in other cells. Therefore they are not classified as organisms, although they show many of the characteristics of living things.

Other classification systems

Classical taxonomy is one of several systems used to classify living things. Another significant method is **cladistic classification**.

Cladistic classification

As with classical taxonomy, this system places groups into larger groups (called **clades**) based on shared characteristics. But clades also include the ancestral forms and all their descendants.

* **Cell wall**, 10; **Cellulose**, 103 (**Roughage**); **Exoskeleton**, 38; **Incisors**, 57; **Larva**, 49; **Multicellular**, 10; **Nucleic acids**, 96; **Nucleus**, 10; **Oviparous**, 46; **Photosynthesis**, 26; **Unicellular**, 10.

INFORMAL GROUP TERMS

Listed here are the main terms used to group living things together according to their general life styles (i.e. their ecological similarities – see also page 9). These are informal terms, as opposed to the formal terms of the classification charts (pages 112-113).

Plants

Xerophytes
Plants which can survive long periods of time without water, e.g. cacti.

Hydrophytes
Plants which grow in water or very wet soil, e.g. reeds.

Mesophytes
Plants which grow under average conditions of moisture.

Halophytes
Plants which can withstand very salty conditions, e.g. sea pinks.

Lithophytes
Plants which grow on rock, e.g. some mosses.

Epiphytes
Plants which grow on other plants, but only to use them for support, not to feed off them, e.g. some mosses.

Saprophytes
Plants which live on decaying plants or animals, feed off them, but are not the agents of their death, e.g. some fungi.

Animals

Predators
*Animals which kill and eat other animals (their **prey**), e.g. lions. Bird predators, e.g. hawks, are called **raptors**.*

Detritus feeders
Animals which feed on debris from decayed plant or animal matter, e.g. worms.

Scavengers
*Large **detritus feeders**, e.g. hyenas, which feed only on dead flesh (animal matter).*

Territorial
*Holding and defending a **territory** (an area of land or water) either singly or in groups, e.g. many fish, birds and mammals. This is usually linked with attracting a mate and breeding.*

Abyssal
Living at great depths in the sea, e.g. oarfish and gulper eels.

Demersal
Living at the bottom of a lake or the shallow sea, e.g. angler fish and prawns.

Sedentary
Staying mostly in one place (but not permanently attached), e.g. sea anemone.

Nocturnal
Active at night and sleeping during the day, e.g. owls and bats.

Plants and animals

Insectivores
Specialized organisms which eat mostly insects, e.g. pitcher plants (which trap and digest insects) and hedgehogs.

Parasites
*Plants or animals which live in or on other living plants or animals (the **hosts**), and feed off them, e.g. mistletoe and fleas. Not all are harmful to the host.*

Mutualists
*A pair of living things which associate closely with each other and derive mutual benefit from such close existence (**mutualism**). **Lichens**, normally found on bare rock, are an example. Each is in fact two plants (a fungus and an alga). The alga produces food (by **photosynthesis***) for the fungus (which otherwise could not live on bare rock). The fungus uses its fine threads to hold the moisture the alga needs.*

Commensals
*A pair of living things which associate closely with each other, one deriving benefit without affecting the other. One type of worm, for instance, is very often found in the same shell as a hermit crab. One of the commonest examples of **commensalism** is the existence of house mice wherever there are humans.*

Social or **colonial**
*Living together in groups. The two terms are synonymous in the case of plants, and refer to those which grow in clusters. In the case of animals, there is a difference of numbers between the terms. Lions, for example, are social, but their groups (**prides**) are not large enough to be called colonies. With true colonial animals, there is also a great difference in the level of interdependence between colony members. In a gannet colony, for example, this is relatively low (they only live close together because there is safety in numbers). In an ant colony, by contrast, different groups (**castes**) have very different jobs, e.g. gathering food or guarding the colony, so each member relies heavily on others. The highest level of colonial interdependence is shown by the tiny, physically inseparable, single-celled organisms which form one living mass, e.g. a sponge.*

Sessile
In the case of animals, this term refers to those which are not free to move around, i.e. they are permanently fixed to the ground or other solid object, e.g. barnacles. With plants, it describes those without stalks, e.g. stemless thistles.

Pelagic
*Living in the main body of a lake or the sea, as opposed to at the bottom or at great depths. Pelagic creatures range from tiny **plankton** through medium-sized fish and sharks to very large whales. The medium-sized and large ones are all animals, and are called **nekton** (from the Greek for "swimming thing"), as they swim.*

Plankton
*Aquatic animals and plants, vast numbers of which drift in lakes and seas, normally near the surface (plant plankton is **phytoplankton**, and animal plankton is **zooplankton**). Plankton is the food of many fish and whales and is thus vital to the ecological balance (**food chains***) of the sea. Most are small.*

Littoral
Living at the bottom of a lake or the sea near the shore, e.g. crabs and seaweed.

Benthos
*All **abyssal**, **demersal** and **littoral** plants and animals, i.e. all those which live in, on or near the bottom of lakes or seas.*

** **Food chains**, 6 (**Food web**); **Photosynthesis**, 26.*

INDEX

The page numbers listed in the index are of three different types. Those printed in bold type (e.g. **92**) indicate in each case where the main definition(s) of a word (or words) can be found. Those in lighter type (e.g. 92) refer to supplementary entries. Page numbers printed in italics (e.g. *92*) indicate pages where a word (or words) can be found as a small print label to a picture. If a page number is followed by a word in brackets, it means that the indexed word can be found inside the text of the definition indicated. If it is followed by (**I**), the indexed word can be found in the introductory text on the page given. Bracketed singulars, plurals, symbols and formulae are given where relevant after indexed words. Synonyms are indicated by the word "see", or by an oblique stroke (/) if the synonyms fall together alphabetically.

ACKNOWLEDGEMENTS

Cover designer: Stephen Moncrieff
Photoshop operator: Fiona Johnson
Additional text by Paul Dowswell
Additional design by Nerissa Davies

Additional illustrations by:

Simone Abel, Dave Ashby, Mike Atkinson, Craig Austin (The Garden Studio), Graham Austin, Bob Bampton (The Garden Studio), John Barber, Amanda Barlow, David Baxter, Andrew Beckett, Joyce Bee, Stephen Bennett, Roland Berry, Andrzej Bielecki, Gary Bines, Derick Bown, Isabel Bowring, Trevor Boyer, Wendy Bramall (Artist Partners), Derek Brazell, John Brettoner, Paul Brooks (John Martin Artists), Peter Bull, Mark Burgess, Hilary Burn, Andy Burton, Liz Butler, Martin Camm, Lynn Chadwick, Peter Chesterton, Sydney Cornford, Dan Courtney, Frankie Coventry (Artist Partners), Patrick Cox, Christine Darter, Sarah De Ath (Linden Artists), Kevin Dean, Peter Dennis, Richard Draper, Brian Edwards, Michelle Emblem (Middletons), Caroline Ewen, Sandra Fernandez, James Field, Denise Finney, Don Forrest, Sarah Fox-Davies, John Francis, Mark Franklin, Nigel Frey, Judy Friedlander, Sheila Galbraith, Peter Geissler, Nick Gibbard, William Giles, Mick Gillah, Victoria Goaman, David Goldston, Peter Goodwin, Victoria Gordon, Jeremy Gower, Terri Gower, Miranda Gray, Terry Hadler, Edwina Hannam, Alan Harris, Brenda Haw, Tim Hayward, Nicholas Hewetson, Philip Hood, Chris Howell-Jones, Christine Howes, Carol Hughes (John Martin Artists), David Hurrell (Middletons), Roy Hutchison (Middletons), Ian Jackson, Elaine Keenan, Roger Kent, Aziz Khan, Colin King, Deborah King, Steven Kirk, Jonathan Langley, Richard Lewington (The Garden Studio), Jason Lewis, Ken Lilly, Steve Lings (Linden Artists), Mick Loates (The Garden Studio), Rachel Lockwood, Kevin Lyles, Chris Lyon, Kevin Maddison, Janos Marffy, Andy Martin, Josephine Martin, Nick May, Rob McCaig, Joseph McEwan, David McGrail, Malcolm McGregor, Doreen McGuinness, Dee McLean (Linden Artists), Richard Millington, Annabel Milne, Sean Milne, David More (Linden Artists), Dee Morgan, Robert Morton (Linden Artists), David Nash, Susan Neale, Louise Nevett, Martin Newton, Barbara Nicholson, Louise Nixon, David Nockels (The Garden Studio), Richard Orr, Steve Page, David Palmer, Patti Pearce, Justine Peek, Liz Pepperell (The Garden Studio), Julia Piper, Gillian Platt (The Garden Studio), Maurice Pledger, Cynthia Pow (Middletons), Russell Punter, David Quinn, Charles Raymond (Virgil Pomfret Agency), Barry Raynor, Phillip Richardson, Jim Robins, Michael Roffe, Michelle Ross, Mike Saunders (Tudor Art), John Scorey, Coral Sealey, John Shackell, Chris Shields (Wilcock Riley), John Sibbick (John Martin Artists), Penny Simon, Gwen Simpson, Annabel Spencerley, Peter Stebbing, Sue Stitt, Roger Stewart, Ralph Stobart, Alan Suttie, John Thompson-Steinkrauss (John Martin Artists), Sam Thompson, Stuart Trotter, Joyce Tuhill, Sally Voke (Middletons), Sue Walliker, Robert Walster, David Watson, Ross Watton, Phil Weare, Wigwam Publishing Services, Sean Wilkinson, Adrian Williams, Adam Willis, Roy Wiltshire, Ann Winterbotham, Gerald Wood, James Woods (Middletons), David Wright (Jillian Burgess), John Yates.

Photograph credits:

Cover (clockwise from top left): © Alfred Pasieka / Science Photo Library; © AJ Photo / Science Photo Library; © David Scharf / Science Photo Library; © Andrew Syred / Science Photo Library.

This edition first published in 2006 by Usborne Publishing Ltd, Usborne House, 83-85 Saffron Hill, London EC1N 8RT, England. www.usborne.com
Copyright © Usborne Publishing Ltd, 2006, 2000, 1986.
The name Usborne and the devices 🌐 ♆ are Trade Marks of Usborne Publishing Ltd.